AN ORIGINAL THEORY OF THE UNIVERSE

HISTORY OF SCIENCE LIBRARY

Editor: MICHAEL A. HOSKIN
Lecturer in the History of Science, Cambridge University

THE ENGLISH PARACELSIANS
Allen G. Debus
Professor of the History of Science, University of Chicago

WILLIAM HERSCHEL AND THE CONSTRUCTION OF THE HEAVENS
M. A. Hoskin
Lecturer in the History of Science, Cambridge University

A HISTORY OF THE THEORIES OF RAIN
W. E. Knowles Middleton

THE ORIGINS OF CHEMISTRY
R. P. Multhauf
Director of the Museum of History and Technology at the Smithsonian Institution

THEORIES OF LIGHT FROM DESCARTES TO NEWTON
A. I. Sabra
Reader in the History of the Classical Tradition, The Warburg Institute

MEDICINE IN MEDIEVAL ENGLAND
C. H. Talbot
Research Fellow of the Wellcome Institute of the History of Medicine, London

THE EARTH IN DECAY
A HISTORY OF BRITISH GEOMORPHOLOGY, 1578–1878
G. L. Davies
Lecturer in the Department of Geography, Trinity College, University of Dublin

THE CONFLICT BETWEEN ATOMISM AND CONSERVATION THEORY 1644–1860
W. L. Scott
*Professorial Lecturer, School of Business Administration and Department
of Mathematics and Science, American University*

THE ROAD TO MEDICAL ENLIGHTENMENT 1650–1695
L. S. King
Professorial Lecturer in the History of Medicine, University of Chicago

WILLIAM HARVEY
AND THE CIRCULATION OF THE BLOOD
Gweneth Whitteridge
Lecturer in the History of Medicine, University of Edinburgh

THE SCIENTIFIC ORIGINS OF NATIONAL SOCIALISM
SOCIAL DARWINISM IN ERNST HAECKEL AND THE GERMAN MONIST LEAGUE
Daniel Gasman
Assistant Professor of History, John Jay College, City University of New York

HISTORY OF SCIENCE LIBRARY: PRIMARY SOURCES

VEGETABLE STATICKS
Stephen Hales
Foreword by M. A. Hoskin
Lecturer in the History of Science, Cambridge University

SCIENCE AND EDUCATION IN THE SEVENTEENTH CENTURY
THE WEBSTER-WARD DEBATE
Allen G. Debus
Professor of the History of Science, University of Chicago

FUNDAMENTA MEDICINAE
Friedrich Hoffmann
Translated and introduced by L. S. King
Professorial Lecturer in the History of Medicine, University of Chicago

THE PNEUMATICS OF HERO OF ALEXANDRIA
Introduced by Marie Boas Hall
Reader in History of Science and Technology, Imperial College

Thomas Wright *of Durham*

AN ORIGINAL THEORY
OR NEW HYPOTHESIS
OF THE UNIVERSE

1750

A facsimile reprint together with the first publication of
A THEORY OF THE UNIVERSE
1734

Introduction and Transcription by
MICHAEL A HOSKIN

MACDONALD· LONDON
AND AMERICAN ELSEVIER INC. · NEW YORK

ASTRONOMY

© Michael A. Hoskin 1971
An Original Theory of the Universe first published 1750
This edition first published 1971

Sole distributors for the United States and Dependencies
American Elsevier Publishing Company, Inc.
52 Vanderbilt Avenue
New York N.Y. 10017

Sole distributors for the British Isles and Commonwealth
Macdonald & Co. (Publishers) Ltd
49–50 Poland Street
London W1A 2LG

All remaining areas
Elsevier Publishing Company
P.O. Box 211
Jan van Galenstraat 335
Amsterdam
The Netherlands

Library of Congress Catalog Card Number 70–139573
Standard Book Numbers
British SBN 356 03515 8
American SBN 444 19612 9

Printed in Great Britain by
Balding and Mansell
London and Wisbech.

CONTENTS

LIST OF PLATES
APPEARING IN THE INTRODUCTION

INTRODUCTION
TO THE MACDONALD ELSEVIER EDITION
BY MICHAEL A HOSKIN

THOMAS WRIGHT 'of Durham' was born at Byers Green, near
Durham City in the north of England, on 22 September 1711, the
son of a yeoman and carpenter.[1] His turbulent boyhood and adolescence
were in marked contrast to the serene and ordered adult life that was to
follow. His early schooling was cut short 'by a very great Impediment of
Speach' and he was, he tells us, 'Very wild & much adicted [to] Sport'.
At the age of 13 he was apprenticed to a clock and watch maker, and
during this time he was 'Very much given to yᵉ Amusement of Drawing,
Planning of Maps and Buildings', developing talents that were to stand
him in good stead in later life.

Wright stayed with his master four years, by which time the interest in
mathematics and astronomy which had been aroused by his childhood
teacher, 'a wrgt good Accomptant and an Astronomer', had become a
dominating passion. Encouraged by his mother, Wright was spending all
his leisure hours studying astronomy, with such total dedication that
'Father, by ill advice, think him mad. Burn all the Books he can get and
endevour to prevent Study.' But soon afterwards, in September 1729, all
hope of study had to be abandoned when Wright became involved in a
scandal and 'Great Disturbances hapn'd and his Master and Mrs not

believing as they should was at last forc'd to run forth'.

After much difficulty Wright managed to reach home, and soon was released from his master by Justices of the Peace. He tried to get work, but without success, so he studied navigation, and wrote 'many things for his own Improvement in astronomy—Geography ec [etcetera]'. He also notes, significantly, 'Reflecting almost upon every object, conseive may find Ideas of ye Deaty and Creation'; and, as we shall see, the task of producing an integrated picture of natural and supernatural, of creation and Creator, was to be the dominating theme of all Wright's cosmological works.

Wright now thought of going to sea, but a storm during his first voyage convinced him that his vocation lay on dry land. He therefore set up a mathematical school at Sunderland in Yorkshire, fell in love with a clergyman's daughter and planned a secret marriage; 'but by an Intelligence from ye B'pp's office at Durham, his Marage was Prevented and Miss Lock'd up'. In his despair Wright headed for London and arranged a passage to the West Indies, but 'being found out by a Friend of my Father's, his going is Prevented'. He then took work in London with makers of mathematical instruments, but after a while his long-suffering father sent money for his journey back north, and Wright once more set up school in Sunderland. There, despite efforts on the part of rivals to have him expelled from the town, he taught navigation to seamen with great success.

With the end of the winter and the departure of the seamen Wright found himself in the spring of 1731 with time on his hands, and he began by planning 'a General Representation of Euclid's Elements in one Large Sheet: and the Doctrine of Plain and Spherical Trigonometry all at one View, on an other'. Such sheets, or 'schemes', were already a feature of the work of other popular teachers of science,[2] but in astronomy Wright was to carry the 'all at one view' technique to the limit and beyond. On p. 1 of the present book Wright speaks of 'a Section of the Creation, eighteen Feet long and one broad' and another 'nine Feet long and six broad'; and, in addition to the sheets of more manageable size which he published on the solar eclipse of 1737 and the comet of 1742, we read in his *Journal* that in 1736 he completed 'his invention of the Theory of Existence and Represent ye Hypotheses in a Section of the Creation 16 feet Long' and the following summer made in brass 'a system of yᵉ Planetary Bodies in true Proportion Equal to a Radius of 190 feet'. In 1737 he drew 'a great many Demonstratory Schemes in Astronomy and Project his Physical and Mathematical Elements', and in his *Journal* he comments: 'His invention at this time run to fast for Execution'. But the most successful scheme was still to come: a chart covering 24 square feet and accompanied by a 'clavis' or key in the form of a substantial volume, *Clavis Cœlestis*, published in 1742.[3]

In addition to planning geometrical and trigonometrical sheets, Wright spent part of 1731 computing and compiling an almanac for the following year, but now his inexperience showed itself: he had allowed insufficient time for the printing, and he had no list of subscribers. The following year he had both time and subscribers to spare, but still had much to learn about the trustworthiness of publishers and engravers. The almanacs were late, and his money 'falling short'. But meanwhile his evident abilities and energy were gaining him influential friends. Wright was a man of lively and versatile intelligence, limited in formal education but full of confidence and willing to turn his mind in any profitable and congenial direction. As an author and teacher of navigation he could no doubt scrape a living; but he realised that as a tutor and consultant to the aristocracy he would share a spacious way of life otherwise denied to him. He had begun modestly enough: in 1731, 'Survey Mr. Aseley's estate at Newfield and make a Map of it.' By autumn 1733 the tide was clearly turning in his favour: 'The Rev. Mr. Newcome introduce to yᵉ Earl of Scarbro', who invites him up to London and Promis's his Patroniseation and Recommendation . . . The Earl of Scarboro' Recommends him to yᵉ Lords of yᵉ Admiralty who give their approbation for Publishing his Pannauticon by Subscription . . . The Earl of Scarboro' obtains leave of the King for him to Dedicate it to His Majesty. Procures him his Royal Highness the Prince for a Subscriber and Recommends him to yᵉ Earls of Ila and Pembroke.'

For many years Wright made a comfortable and agreeable living in high society, holding public and private classes, surveying and advising on the estates of the rich, and publishing a number of books and illustrated sheets. Typical are the entries in his *Journal* for spring 1739: 'Project of a Large Horizontal Dial for Lord Viscount Middleton. Go with his

Lordship to his seat in Surry; stay a month with him. Spend his time extremely agreeable. Lady Midleton Lady Charlotte and Lady Mary Capel study yᵉ use of yᵉ Globes.'

Such was Wright's reputation that in 1742 he was offered £300 a year to become professor of navigation at the Imperial Academy at St Petersburg. He preferred to remain in England, but the note in his *Journal* that 'His Proposals of £500 were sent to Russia' suggests that he was not without his price. Four years later Wright went to Ireland for several months to assemble drawings for his most successful book, *Louthiana*,[4] a description of the antiquities of the County of Louth published in 1748 and again in 1758. But a sequel to *Louthiana* and a volume on the antiquities of England were to remain in manuscript.

In the 1750s Wright's thoughts began to turn once more to his birthplace, no doubt (as Edward Hughes suggests)[5] because nearly all his noble patrons had died off. In 1756 he laid the foundation of a house at Byers Green, and six years later he retired there to 'prosicute my Studies'; many of the Wright manuscripts auctioned by Sotheby's on 19 July 1966, and now in Durham University, belong to this period of retirement. He died in 1786.

Apart from *Louthiana* and a work on *Universal Architecture* which appeared in 1755, Wright's published books all relate to astronomy. The earliest, entitled *Clavis Pannautici, or a Key to the Universal Mariner's Magazine*, is a slim booklet of which Wright's own copy survives at Durham University. It was published in London in 1734, a year after Wright's *Journal* remarks, 'Invents and Completes His Pannauticon, the Universal Mariner's Magazine'. The *Pannauticon* itself has been described as 'a treatise on navigation'[6] and the *Clavis* regarded as 'an introduction to this presumably large volume', and in consequence the absence of any copies of the *Pannauticon* has seemed surprising. But the *Pannauticon* is surely not a book at all; just as the *Clavis Cælestis* was to be 'the explication of a diagram',[7] so the *Clavis Pannautici* is the guide to the use of an instrument. On p.19 of this guide we read: 'The Instrument consists of four Schemes, one fixt, and three moveable, and delineated in the following Manner . . .', and on p.18 Wright speaks of 'large Schemes eighteen inches Diameter, delineated with divers Circles'. Advertisements on two charts which Wright published for the solar eclipse of 18 February

1736/7 provide the necessary identification. They both state that 'Mr Wright has lately publish'd by Subscription the *Perpetual Pannauticon or Universal Mariners Magazine*, being a Mathematical Instrument', and the chart showing the passage of the penumbra over Scotland specifies that the instrument is '18 inches Diam^r describing the Lunar Theory and motion of the Tides'.[8]

Wright's first major work was destined to remain in manuscript. It is a beautifully executed folio volume now at Durham University Library, and is dated 1737. The title runs: '*The Universal Vicissitude of Seasons* exhibiting by inspection at one view, the various rising and setting of the Sun to all parts of the World, with the hour and minute of day-break, length of day, night and twilight etc every day in the year. Illustrated with near 200 copper-plate impressions upon folio paper, all original diagrams curiously delineated, besides the decoration of above 500 various representations of the Sun, and other emblematic ornaments; together with proper tables and demonstratory schemes.' The same library also owns another elaborate unpublished manuscript, undated, in which Wright repeatedly divides and subdivides all knowledge, the divisions being invariably into three parts and resulting eventually in 2187 sections. A similar obsession with triads is to be found in other manuscripts of Wright but nowhere with the thoroughness of this *Pansophia*: a reminder, if one were necessary, that we are dealing with no twentieth-century scientist.

The second of Wright's books to reach publication appeared in 1740: *The Use of Globes*, an introduction to the celestial and terrestrial globes of John Senex with whom Wright had long had business associations and who was to die later the same year. This little book was followed in 1742 by the vast chart (of which only two complete copies are known to survive) and the accompanying key or *Clavis Cœlestis*. The chief merit of this work is, as the author says, his 'Method of treating it, *viz.* clearly and plainly',[9] but there are two features worthy of particular mention. First, Wright points out in words and in drawings how appearances vary with the position of the observer. For example, he tells us (after the style of Christiaan Huygens and more especially James Gregory) that

To *Mercury*, the *Sun* and *Venus* are the only two great Bodies of the Universe. He views *Venus* and all the rest of the Planets, as we do *Saturn*, *Jupiter* and *Mars*;

but *Venus* shines upon him with great Lustre, and 'tis probable, her great Light in opposition to the Sun, serves him instead of a Moon.[10]

Second, and more significantly, Wright does this not only for the planets but also for the stars, which he believes are equal in size and distributed irregularly:

The Stars are looked upon by all modern Astronomers, to be no other than great Globes of Fire like the Sun, promiscuously distributed through the Mundane Space. . . . Their various apparent Magnitudes are intirely the Effect of an unequal Distance as will manifestly appear from the Scheme of their Disposition . . . .[11]

And in the scheme or figure he shows how double stars and even clusters may be optical effects due to the particular location of the observer.

Wright's *Journal* ends in 1746, so our knowledge of his activities in the months immediately preceding the publication in 1750 of his *Original Theory of the Universe* is slight. We do, however, possess a copy of the prospectus of the work.[12] This contains some ten plates and a table of contents, together with a title page:

Now in the Press, and shortly will be publish'd by Subscription, *An Original Theory* . . . . *Half a guinea* to be paid at subscribing, and *Five Shillings* on Delivery of the Book in Sheets, which will be early in the Spring; and if any remain above what are subscribed for, they will be sold at *One Guinea*. Subscriptions are taken in for the Author, at Mr. Pattison's in *Oxenden-Street*, near the *Hay-Market*, and at Mr. Chapelle's, in *Grosvenor-Street*.

Apart from a late work on longitude, the *Original Theory* was to be Wright's last published book in astronomy, and it is his chief claim to a place in the history of the subject. In contrast to the concisely-factual approach of much of the *Clavis Cœlestis*, the writing is diffuse and theologically-orientated; but, as Wright insists repeatedly, it is directed to one central problem, the explanation of the Milky Way. This intriguing phenomenon had not been mentioned in the *Clavis Cœlestis*, but in writing the *Original Theory* he admits that 'This luminous Circle has often engrossed my Thoughts, and of late has taken up all my idle Hours' (p. 37). Can we recover something of the train of his thought as he worried his way to a solution of the mystery?

Wright himself provides the key on p.1 of the *Original Theory*, where he tells us that 'The Hypothesis upon which this new Astronomy is founded, and now reduced into a regular System, was the result of my Astronomical Studies full fifteen years ago'. In a footnote he is more specific: 'The first Scheme of this Hypothesis was plann'd in the Year 1734, representing in a Section of the Creation, eighteen Feet long and one broad, several thousand Worlds and Systems, and a great Number of emblematical Figures. . . .' From this two significant facts emerge: first, in 1734 Wright was producing a vast visual aid in the form of a flat, two-dimensional drawing, and this, as we might expect, showed a flat, two-dimensional object, a plane section of the Creation; second, from the talk of emblematical figures, we learn that Wright was by no means engaged on pure science.

Fortunately, although the scheme itself seems not to have survived, a pair of documents related to it have recently come to light and are published for the first time in this volume. With their aid we shall now essay a tentative reconstruction of the development of Wright's thinking about the Milky Way. But as we do this we shall realise that in the general context of Wright's life-work the explanation of the Milky Way is a side-issue, temporarily given prominence in 1750; and that the attempts to give an integrated picture of the moral and physical universe, so often deplored by critics of the *Original Theory*, represent the permanent elements in his cosmological thinking. We shall then be less surprised to find that in later life these attempts led him to a radical reappraisal in the course of which, in a mere sentence or two, he offers an utterly different explanation for the Milky Way.[13]

The first document of the pair was once headed 'The Elements of Existence or a Theory of Eternity', but 'Theory of Eternity' has been altered to 'Theory of the Universe'. It is so brief as to be little more than a descriptive title to the scheme, which it says is

a section of ye Univers . . . comprehending first ye Paradise of imortal spirits in there several Degrees of Glory, surounding the Sacred Throne of Omnipotence. Secondly, the Gulfe of Time or Region of Mortality, in which all sensible beings such as ye planetary bodies are imagind to circumvolve in all maner of direction round the Devine Presence, or ye Eternale Eye of Providence. Thirdly, the shades of Darkness & Dispare supposed to be the Desolate Regions of yᵉ Damnᵈ.

On the next page we find a note: 'Wrote in yᵉ year 1734: the author being then 22 years old.'

The second document is a much longer 'explanation' of this *Theory of the Universe* and reads like the text of a lecture discussing the scheme. The importance of the lecture (if such it be) is indicated by the note added at the end:

This Juvenil Performance was the Produce of the authors Imagination before he had Reap'd any advantages either from Reading or Study, but Prov'd afterwards the foundation of his Theory of yᵉ Univers a much more perfect Work.

But here Wright doth protest too much, for in 1729, when he noted in his Journal 'Conseive may find Ideas of yᵉ Deaty and Creation', his father had recently burned his books because Wright was so madly devoted to the study of astronomy. Before discussing the *Theory of the Universe*, then, let us examine some of the ideas current in books he is likely to have read at this impressionable age.

Among the books on astronomy which enjoyed a wide circulation in English at this time were Christiaan Huygens's *Cosmotheoros* (Hagæ-Comitum, 1698), of which English editions appeared under the title *The Celestial Worlds Discover'd* 1698 and again in 1722; David Gregory's *Astronomiæ Physicæ et Geometricæ Elementa* (Oxford, 1702), published in English as *The Elements of Physical and Geometrical Astronomy* in 1715 and 1726; William Whiston's *Prælectiones Astronomicæ* (Cambridge, 1707), translated as *Astronomical Lectures* and published in 1715 and 1728; John Keill's lectures *Introductio ad Veram Astronomiam* (Oxford, 1718), translated as *An Introduction to the True Astronomy* and published in 1721, 1730 and subsequently; and William Derham's *Astrotheology: or, a Demonstration of the Being and Attributes of God, from a Survey of the Heavens*, published in London in 1715 and many times subsequently. To these we should add Whiston's *Astronomical Principles of Religion* (London, 1717 and 1725), on which Wright drew extensively when writing *Clavis Cælestis*, and two works carrying all the prestige of Isaac Newton: the English translations of his *Principia* (London, 1729) and the much more manageable *System of the World* (London, 1728).

From these works Wright would learn that the stars are self-luminous like the Sun and probably have their own planets and comets. Most authors (but not Whiston) would go further and say that the stars are similar in size to the Sun, and some set out methods (due to James Gregory and Huygens) of estimating the distances of stars on the assumption that they appear fainter than the Sun only because they are more distant.

Most of the books would tell Wright that the visible stars are only a part of the total creation, and perhaps an altogether insignificant part. But Huygens thinks it possible that 'beyond such a determinate Space [of stars] he [God] has left an infinite Vacuum';[14] and Whiston, in a phrase that could almost have been written by Wright, speaks of 'that unfathomable Abyss of infinite Extramundane Space'.[15] As to the arrangement of the stars, suggestions that they are regularly spaced out are perhaps less significant than the view of Whiston and Derham that the stars appear disordered only because of our position as observers. For Whiston 'it is very rational to conclude, that some regular Order hath Place also amongst the Fixed Stars. There may be a certain orderly and harmonious Disposition of the Fixed Stars amongst themselves, when they are beheld from some other proper Place, altho' that Order appears not when they are seen from this Earth.'[16] To Derham the stars

look to us, who can have no regular prospect of their positions, as if placed without any Order : like as we should judge of an Army of orderly, well disciplined Soldiers, at a distance, which would appear to us in a confused manner, until we came near, and had a regular prospect of them, which we should then find to stand well in rank and file.[17]

But perhaps the most striking passage for our present purpose occurs when Whiston, writing before Halley's discovery that certain stars had a motion of their own,[18] forecasts the danger of destruction of the 'Grand System' of stars:

. . . since withal, the Sun and Fixed Stars do not revolve about one another, or about any common Center of Gravity . . . it follows, that the several Systems, with their several Fixed Stars or Suns, do naturally and constantly, unless a Miraculous Power interposes to hinder it, approach nearer and nearer to the common Center of all their Gravity ; and that in a sufficient Number of Years, they will actually meet in the same common Center, to the utter Destruction of the whole Universe.[19]

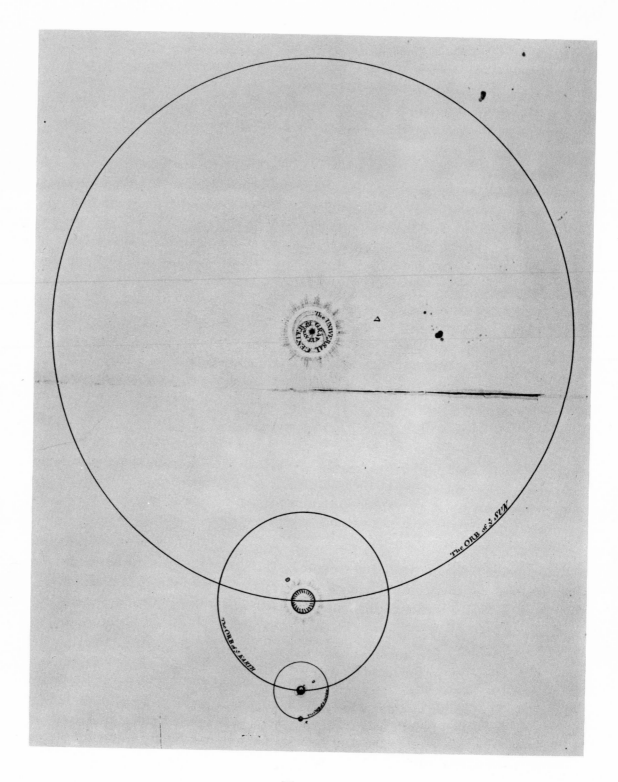

Plate I

Sketch by Wright, presumably for Plate XXI of *An Original Theory*, showing the hierarchy of orbital motions: the Moon in orbit about the Earth, the Earth about the Sun, and the Sun about 'The Universal Center of Gravitation'.

It seems a natural reaction of someone reading this after Halley's discovery, being familiar also with the hierarchical division of Sun, 'primary' planets and 'secondary' planets, and sharing Whiston's concern to integrate the religious and astronomical views of the universe, to suggest that stars of the Grand System *are* in motion about their common centre of gravity, and that this is why the System does not collapse into its centre. Such an orbital motion about the centre is fundamental to Wright's thought in 1734 as in 1750, motion both of the stars in general and of the Sun in particular. Plate 1, a previously unpublished sketch[20] presumably for Plate XXI of the *Original Theory*, clearly indicates the hierarchy of orbital motions and the path of the Sun about 'The Universal Center of Gravitation'. The 1734 description cited above of the 'Gulfe ... in which all sensible beings such as yᵉ planetary bodies are imagined to circumvolve in all maner of direction round the Devine Presence' is developed at length in the sixth letter of the *Original Theory*. There Wright asks the reader (p.51) to grant that 'all the Stars are, or may be in Motion'; he cites Halley's paper (pp.53-54), and argues that it is by these motions that the stars are prevented 'from rushing all together, by the common universal Law of Gravity' (p.57). But he was more successful in representing the orbital motions of the stars in sketches hitherto unpublished (Plates 2-4) than in the figures of the *Original Theory*.[21]

There was much else in Whiston's *Astronomical Principles of Religion* to appeal to a man of Wright's temperament: talk of 'the noblest or invisible Parts of the Creation', of 'this System of the Universe' being 'God's great House, or Family, or Kingdom', and of the location of Heaven, Hell, and the various spirits.[22] For Whiston, the Devil and his angels, together with wicked men, will one day be transferred from their prison within the Earth to a comet, there to be tormented in the sight of the Blessed who will inhabit our atmosphere. Only slightly less crude is the famous treatise by Tobias Swinden (cited by Derham), *An Enquiry into the Nature and Place of Hell*.[23] According to Swinden, Hell is in the Sun, which is the centre of the 'orbicular' Creation; around the Sun moves the Earth, and around and outside the Earth are the vast numbers of stars, beyond which is the Throne of God.[24]

Wright's own picture concentrates on Hell as a place of darkness rather than of heat, and it lies outside the circle of stars. In this he may well

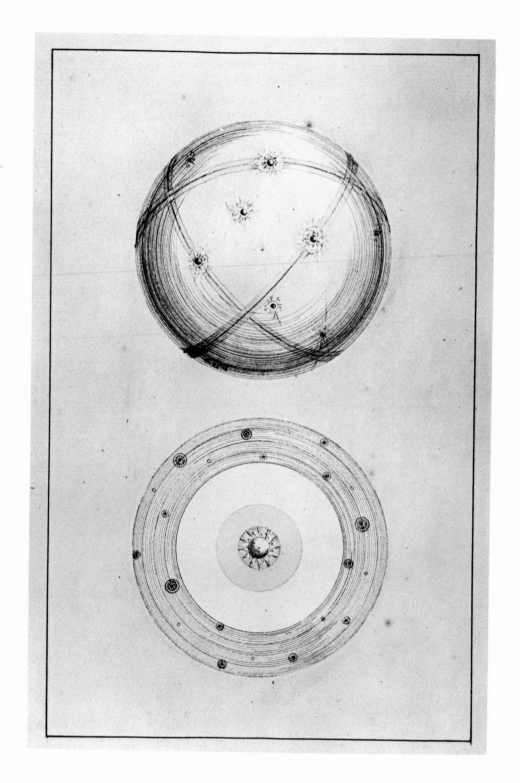

Plate 2

Sketches by Wright, presumably for Plates XXIV and XXV of
An Original Theory, showing a spherical shell of stars which are
orbitting in various directions about the centre.

Plate 3
Further sketches by Wright showing a spherical shell of stars which
are orbitting in various directions about the centre; the motions of
the stars are better represented here than in the published plates
of *An Original Theory*.

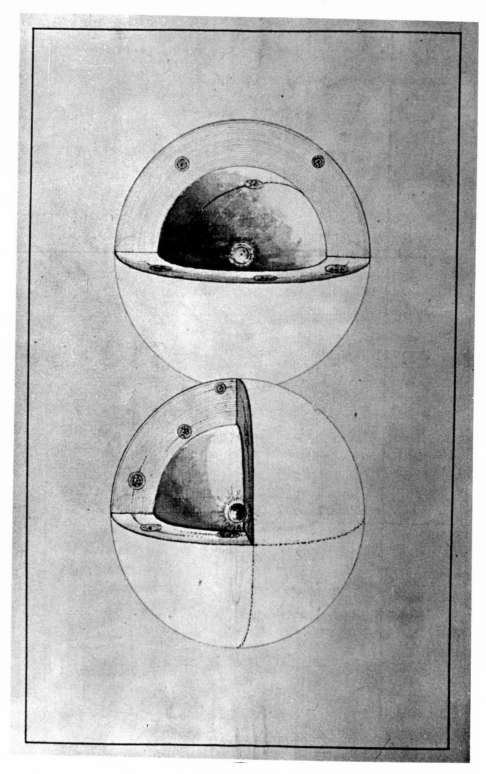

Plate 4
Yet more attempts by Wright to represent the orbital motions
of stars.

have been influenced by Robert Fludd,[25] or by the German mystic Jacob Boehme or Behmen whose writings were so popular in early-eighteenth-century London[26] and which contain many elements found in Wright's works: for example, Boehme's *XL Questions Concerning the Soule* tells us of the 'Eye of Eternity', the 'Angelicall World . . . Court of Angels or Princely Thrones in the Liberty of the Divine Majesty', and the 'Abysse of the Darke World';[27] and elsewhere Boehme presents us with the central Eye surrounded by stars which in turn are enclosed by the serpent near to biting its own tail.[28] But Wright's immediate source may well be a more recent work as yet unidentified, in which these ideas are assimilated to the recognition of the Sun as one of the stars.

In this early lecture on his *Theory of the Universe*, we find that Wright is already committed to his life's work in cosmology: the reconciliation of the moral and the physical view of the world. Already he believes *the world has a centre in the moral order*, the Sacred Throne, *which he identifies with the centre in the physical order*, the gravitational centre and source of the laws of nature; and already *the very existence of such a centre is for him a guarantee that both morally and physically the creation is arranged symmetrically around this centre*. We might here interpose the question, Is such a reconciliation worth attempting, in view of the fact that our observational knowledge of the physical order is confined to one little corner of creation? Yes, precisely because of the symmetry about the centre; for, as Wright will argue at length in the second letter of the *Original Theory*, 'from a very small part of orbicular things, we are able to determine the form and direction of the whole' (p.11).[28a]

Wright was to express these fundamental tenets in a different way in each of his three cosmological writings. In the 1734 *Theory of the Universe* he does it in terms of two spheres, both centred on the Sacred Throne. Inside the smaller sphere is the Region of Philisity or Heaven; outside the larger sphere is the Region of Punishment or Hell, the outer darkness; while in the spherical shell lying between the two spheres is the Region of Probation, and here the solar system, like the other 'miriads of systems in all manner of dispositions', circles about the Sacred Throne.

To add force to his message Wright proceeds to take his audience on a tour of these various regions, after the manner of Keill, who had invited his readers to 'suppose we have the Power of Travelling every where, thro' the Immense Regions of indefinite Space'.[29] For example, some parts of Hell we find are 'faintly enlightened' by the light from our stars, while from other, nearer parts we can see the stars (no doubt by analogy with nebulae) as forming 'a bright and glorious creation'.

Now Wright is very far from being the devout but critical scientist eager, in Whiston's phrase, to compare these two Divine Volumes of revelation and natural knowledge. But to carry conviction he must make some show of relating his overall conception to what we know by observation of our particular corner of the universe, 'demonstrating', as he says in his synopsis, 'ye visible creation to be only a finite view of ye general frame of Nature'. And so, in his scheme (C3 & 4)

upon all sides the principal stars of the visible creation are exhibited in ther natural order as seen from ye Earth by ye naked eye. Those of ye first magnitude nearest to our own system, and the rest proportionable removed according to their respective phenomena. Beyond these are others more remote crowned by a penumbral shaddow such as we call telescopic stars and again without them more, suposd to be at immense distance & by no means perceptible to ye human eye. At a certain distance from ye Sun equal to a vissual ray of ye smallest visible star is a faint circle of light terminating the utmost extent of ye visible creation, in a finite view from ye Earth . . . .

This explanation of the Milky Way, the 'faint circle of light', is momentous. But Wright's attempt to reconcile observational evidence with his overall theological picture is no more than cursory. The plane section represented in his 'scheme' he defines as 'passing thro' the center of ye [S]edes Beatorum & coinciding with that of our Earth or ye solar orbit'(B1), and he also says it is 'parallel to ye Orbis Magnus' (C3). Is then this versatile plane to contain the Sacred Throne or Sedes Beatorum and the Milky Way and the ecliptic? No doubt we must make full allowance for artistic license; but this must not blind us, as it has blinded Wright, to one fundamental error: the Milky Way is unique, whereas a faint circle of light should be seen in any and every plane drawn through the position of an observer located within 'miriads of systems in all manner of dispositions'. In other words, if the stars were scattered around us in all directions then the entire sky would seem ablaze with a faint light.

Whether Wright had realised this by 1742, when his Clavis Cælestis

xxiv

appeared, is doubtful. We would know from internal evidence alone that he had not solved the problem, for he describes the stars as 'promiscuously distributed through the Mundane Space'.[30] But if the Milky Way is the effect of very many distant stars concentrated in certain directions only, the true distribution cannot be promiscuous. By 1750 he had realised this: as he tells us on p. 30 of the *Original Theory*, the stars are 'distributed either promiscuously, or in some regular Order', and his task is to show that the latter alternative is correct.

In the famous middle section of Letter the Seventh, Wright invites us to 'imagine' the stars as scattered in the region bounded by two parallel planes. An observer within this region looking along the layer would see bright near stars, then fainter and more distant stars, and then still fainter and more distant stars, in such numbers that their light would appear to 'form a perfect Zone of Light; this I take to be the real Case, and the true Nature of our *Milky Way*' (p.63). By contrast, the same observer looking *away* from the layer of stars would see only a few bright stars before his gaze reached empty space.

This explanation of the Milky Way as an optical effect due to our immersion in a layer of stars constitutes Wright's chief claim to fame, and the accompanying Plate (XXIII, engraved in error as XXI) is among the most familiar in the history of astronomy. But so far Wright has discussed only a simplified model for the distribution of the *visible* stars, and as he himself immediately says: 'But now to apply this Hypothesis to our present Purpose, and reconcile it to our Ideas of a circular Creation' (p.63). He has still to reconcile this scientific hypothesis that the visible stars form a flattish layer with his theological and moral world picture which by 1750 he had modified and extended but by no means abandoned —a picture in which our star system taken as a whole surrounds its supernatural centre. How was he to do this?

With the unreal clarity that hindsight allows, we see that he could have reasoned somewhat as follows. In 1734 he had drawn a plane section passing through both the supernatural centre and the location of the observer in the solar system. The plane section, on the 1734 world view, could indeed have been any one of the infinitely-many such planes, and yet always it would have contained a milky way. Since the Milky Way is in fact unique and lies in a unique plane passing through the location of the observer, it can actually belong to at most one, and perhaps to none, of these infinitely-many planes.

If it belongs to *one* such plane—that is, if the stars are specially related to a single plane through the supernatural centre—then, according to Wright, our star system *as a whole* is analogous to the rings of Saturn (see Plates XXVIII and XXIX) though the *visible* stars occupy a disk-shaped region in one corner of the ring. On the other hand, if *no* such plane is singled out to contain the Milky Way (and this, with its greater symmetry, clearly appears to be Wright's preferred alternative), then the annulus of stars which lay in the 1734 plane section should have been drawn as too thin to produce a 'faint circle of light' all around the observer. The observer will then see a milky effect in this plane only when he looks tangentially along the annulus. And since the annulus is thin, it follows that the spherical shell of stars of which it is a section is likewise thin. In other words, our star system forms a thin spherical shell surrounding the supernatural centre on all sides; the plane of the Milky Way is the tangent plane to the sphere at the point where the observer is located; and the visible stars are part of a gently-curving spherical layer.

Wright did not find this conception easy to represent, though he devotes four plates to the subject (nos. XXIV–XXVII), and made several other sketches (Plates 2-4). They show that, although Wright will now allow more elements into his universe in the form of multiple or additional systems of stars, on this preferred alternative the 1734 picture persists as essentially the true representation of our own star-system. Nothing would indicate more clearly just how far Wright is from the *modern* conception of our star-system or Galaxy, although it is of course true that his simplified model of the distribution of the *visible* stars has stood the test of time. Strangely, this latter achievement has succeeded almost in obliterating knowledge both of his overall picture of our star-system and of the theological motives and framework to which he was dedicated. As a result not only is Wright often credited with a disk- or lens-like stellar universe, but he has even been described as 'the first astronomer of the "new philosophy" to tackle the problem of the structure of the universe'.

This situation has arisen from a curious set of circumstances. Because of the rarity of the *Original Theory*,[31] Wright's work was known for many years mainly from references in Kant's *Allgemeine Naturgeschichte und*

Theorie des Himmels (1755, although distribution was blighted by the bankruptcy of the publisher).[32] Kant in turn acquired his knowledge of Wright's *Original Theory* from a long review which appeared in 1751 in a Hamburg journal.[33] This account is fair and contains extended quotations, especially of the middle section of the seventh letter, but the two alternative pictures of our complete star system are reduced to a sentence apiece, and the supernatural centre appears to be the centre of creation as a whole and not the centre of our star system. In the inevitable absence of the plates, which show not only that the (local) supernatural centre is in fact in the midst of our stars but also that Wright would have a very great gap between opposite sides of our star system to allow plenty of room for this supernatural centre, Kant not unreasonably thought he was being offered two alternative pictures of our star system, both on a purely natural level. The spherical-shell alternative he felt he could exclude on observational grounds, since he gathered from Maupertuis[34] that nebulae (which Kant believed to be other star systems like our own) appeared elliptiform and not circular. There remained the Saturn's-rings alternative, described in the review by analogy with the solar system, the stars 'all moving in the same way, and not much deviating from the same plane, as the planets in their heliocentric motion do round the solar body'.[35] This analogy greatly appealed to Kant; he was convinced it was sound and that the stars of our system occupy a disk-like region, and he was happy to acknowledge to posterity the debt he imagined he owed to Thomas Wright for this conception.

The next to hit on the same idea was J. H. Lambert, whose *Cosmologische Briefe über die Einrichtung des Weltbaues* was published in Augsburg in 1761. According to a letter he wrote to Kant in 1765 (and there is no reason to doubt its truth), Lambert first outlined his thoughts on the subject in 1749. It was only in 1761, after the composition of the *Briefe*, that he 'was told at Nürnberg that some years previously an Englishman had printed similar thoughts in letters to certain persons, but that he had not had much success, and that the translation of his letters, begun at Nürnberg, had not been completed'.[36]

The concept passed into the mainstream of technical astronomy through the great papers on 'the construction of the heavens' published in the 1780s by William Herschel.[37] Herschel did not, as is sometimes stated, 'prove' the correctness of the view of the Galaxy as disk-shaped, unless by this we mean merely that he confirmed by actual counts the impression we have that more stars lie in the direction of the Milky Way than in other directions. Granted this impression is correct, Herschel's approach was sure to result in a (roughly) disk-shaped Galaxy, since he assumed that the stars are regularly distributed and that his telescope could penetrate to the borders of the Galaxy in every direction. Herschel's triumph was rather to demonstrate the possibilities of stellar statistics and to promote the whole question of the Milky Way beyond the limits of the merely speculative; but it was open to later astronomers to question the validity of his assumptions (as indeed Herschel himself did in later life), and to suggest instead that the Milky Way is what it looks to be: a ring of small stars surrounding a collection of larger stars which includes our Sun.

But did Herschel derive his interest in the problem from Wright, or Kant, or Lambert, or did he hit upon it independently? Probably we shall never know: as a boy in Hanover Herschel had been introduced to astronomy, but he came to England in 1757 as a refugee after Hanover was occupied by the French, and he can hardly have then known of the copies of Kant's work appropriated from the bankrupt publisher; on the other hand he may well have learned of Wright's *Original Theory* in the early 1760s when he travelled extensively in the north-east of England as a music teacher. He makes no mention of Wright, and his copy of the *Original Theory* is little-marked except for a note which could have been written any time after 1781.[38] There are, however, a number of curious ideas common to both the *Original Theory* and Herschel's published papers, which are sufficiently unusual to constitute some evidence that Herschel may have read the *Original Theory* early in his career. For example, Wright's proposal (pp. 55–56) that changes in stellar magnitudes may be caused by movements of stars is echoed in a 1783 draft where Herschel speaks of 'changes in the apparent magnitudes and reciprocal distances of the stars';[39] and Wright's suggestion (p. 45), quickly withdrawn, that stellar distances are directly proportional to their apparent magnitudes as classified by astronomers was an hypothesis used by Herschel for most of his career. Yet both were sufficiently implausible to arouse most strenuous resistance from Nevil Maskelyne when incorporated in papers submitted to the Royal Society.[40]

xxviii

In the nineteenth century Wright continued to be known mainly through the references in Kant, although an undistinguished edition of the *Original Theory* was published without the plates in 1837 in the U.S.A.[41] The relevant plates were again missing from the 'Account of the Speculations of Thomas Wright of Durham' given by Augustus De Morgan in 1848, although (unlike the Hamburg reviewer) he does quote the later parts of Letter the Seventh in full.[42] The text references to Saturn's rings as the analogy for one model of our star system should, one might think, have been clear enough even in the absence of the plates. But De Morgan too makes the fatal transition from the visible stars to our complete star system. Wright, he correctly says, 'gives the explanation of the phænomena of the milky way, as now generally received',[43] but later varies this to 'He gave the theory of the milky way which is now considered as established, contended for what is now called the central sun . . .'.[44] For Kant and his successors this was, as we have seen, an easy transition, encouraged by the partial account of Wright's views that alone was available to them; the failure of De Morgan, and others with access to the *Original Theory*, to recognise the misapprehensions of earlier authors has done justice neither to Wright's own thought nor to Kant's distinctive contribution.[45]

If we now turn our attention from Wright's treatment of the Milky Way and survey the *Original Theory* as a whole, we find that in its overall arrangement, the book is irritatingly diffuse: the blame for much of the misunderstanding with which Wright's work has been surrounded must rest with him. Yet to deplore his religious motives, and to regret the space he devotes to poetry and to history (in which he was greatly indebted to Sherburn's rambling *Sphere of Manilius*)[46] is anachronistic, as the first letter illustrates: a reverential and poetic approach to Nature was common in that more spacious age, and severely technical textbooks of astronomy were out-numbered by works inspired by a more humane outlook.

In Letter the Second, Wright justifies his future use of analogy, as for example when he will assume the stars have their own planetary systems; and he shows how, when we study objects with circular or spherical symmetry, we can reason from the part to the whole. He then with slight excuse interpolates a summary of primitive conceptions of the heavenly spheres.

The third letter gives details of the solar system, and includes the suggestion (p.21) that the orbits of all comets may be equal in area. Here he prints calculations entered in manuscript in his own copy of *Clavis Cælestis*, now in Durham University Library. Wright's suggestion is quoted at length by the Hamburg reviewer and described as 'a very happy remark', but De Morgan, writing a century later and looking for elements of Wright's *Original Theory* still scientifically acceptable, takes the very opposite view: 'This is not a happy conjecture.'[47]

Letter the Fourth is notable, as we have seen, for containing the germ of Wright's insight into the structure of the Milky Way: that the stars are distributed, not at random, but 'in a regular Order' (p.30). Wright's reference in the same letter to 'the *known* Planets' (p.31, italics ours) has been praised as an anticipation of the discovery of Uranus. But he is merely repeating a phrase used by Whiston in 1717 in the *Astronomical Principles* and by himself in 1742 in *Clavis Cælestis*, and the possibility of planets beyond Saturn is elaborated at length by William Wall in 1727 in the second edition of Swinden's *An Enquiry into the Nature and Place of Hell*.[48]

By contrast, a recent critic has with strange perversity compared Wright unfavourably with Kant for advocating, here and elsewhere, the 'queer idea' that the gravity of each star system was limited. We must not forget that at this time there was no proof that gravity did extend beyond the solar system: not until 1767 did John Michell show that multiple stars occur far more often than could be expected if no physical causes were at work;[49] not until 1803 did Herschel give observational evidence that pairs of stars were in orbit around each other under attractive forces, whatever they might be;[50] and it was decades more before these forces were identified as gravitational. But in fact Wright's critic has fastened too easily on a single sentence. Newton (cf. *Original Theory*, p.6) and many subsequent authors had contrasted the effective operation of the Sun's gravitational attraction on planets and comets with the negligible consequences of the attractive forces of other stars. Wright, in spite of his mysticism a very practical man, concentrates on the '*sensible* Sphere of the Sun's Attraction' (p. 31), on 'a Space sufficient to divide or seperate the *sensible* activity of neighbouring Systems' (p. 70, all italics ours), and by so concentrating lapses into occasional ambiguity. His mode of thought is clear from a section of the *Clavis Cælestis*, where he says of two bodies: 'If the Body C had no natural tendency to the Body A, or more properly, if the Body A were removed to an infinite Distance from that of C, so as to render its Virtue of no Effect at C . . .'.[51] In the *Original Theory*, as in the 1734 lecture, Wright has the stars in projectile motion about a centre and it is this that prevents them 'from rushing all together, by the common Universal Law of Gravity' (p. 57).[52]

Letter the Fifth gives Wright the excuse to include some of his most delicate illustrations, based in part on his own modest attainments as an observer. His ideas on the distances separating stars, and on the number of stars visible in the Milky Way, are conservative enough, but, even so, combined they reveal a vastness in the visible creation the excitement of which we can still recapture. Significant for the transformations that overcame Wright's ideas in later life is his account (p. 43) of 'the astonishing Phenomenon of several new Stars', a topic which was to play a crucial role in his *Second Thoughts*. The letter closes with a hint (p. 46) of 'a general Motion of [the stars] round a common Center', and with an attempt to whet our appetite for the future happiness awaiting us among the stars and in 'the final Coalition' in the Abode of the Blessed.

In the sixth letter, Wright argues that the stars are individually in motion, and that it is these projectile forces that prevent the system from collapsing. This analogy with the planets (and with the 'lesser planets' he believed formed Saturn's rings) he continues in Letter the Seventh; and just as the planets appear erratic because of our eccentric position,

we may readily imagine, that nothing but a like excentric Position of the Stars could any way produce such an apparently promiscuous Difference in such otherwise regular Bodies. And that in like manner, as the Planets would, if viewed from the Sun, there may be one Place in the Universe to which their Order and primary Motions must appear most regular and most beautiful. Such a point, I may presume, is not unnatural to be supposed, altho' hitherto we have not been able to produce any absolute Proof of it. See *Plate* XXV [which shows a section of our spherical creation, an annulus of stars 'with the Eye of Providence seated in the Center']. This is the great Order of Nature, which I shall now endeavour to prove and thereby solve the Phænomena of the *Via Lactea*.[53]

This passage alone would be sufficient to demonstrate that Wright plans to integrate his explanation of the visible Milky Way (or, rather, his first simplified model for such an explanation), *which immediately follows*, into his overall conception of the stars as surrounding the supernatural centre; and that it was therefore utterly impossible for him to propose a grind-stone theory of our star system.

We have already discussed how this integration was achieved in the remainder of the letter, and why Kant and subsequent writers failed to realise what Wright was about. The final two letters might have made Wright's position clear, had his readers not been discouraged by their philosophical titles. Letter the Eighth, for example, contains an estimate (p. 73) of the scale of our star system: the shell (or ring?) has a thickness of only some eight million million miles, but an inner diameter of anything from twenty-six to ninety-six million million miles; and a star may take over a million years to orbit around the supernatural centre (p. 75). Letter the Ninth takes it as granted 'that the Creation may be circular or orbicular' (p. 78) and 'that all the Stars may move round one common Center' (p. 79), and enquires into the precise nature of this supernatural centre. In this section, more than anywhere else, Wright's unification of the natural and the supernatural, of efficient causes and final causes, is clearly expressed:

To this common Center of Gravitation, which may be supposed to attract all Vertues, and repel all Vice, all Beings as to Perfection may tend; and from hence all Bodies first derive their Spring of Action, and are directed in their various Motions.

Thus in the *Focus*, or Center of Creation, I would willingly introduce a primitive Fountain, perpetually overflowing with divine Grace, from whence all the Laws of Nature have their Origin . . . .[54]

In the final paragraphs, Wright supposes that 'the endless Immensity is an unlimited Plenum of Creations not unlike the known Universe' (p. 83) and that the nebulae may be just these. He has several times (pp. 41, 42, 65) suggested that not only the Milky Way but other 'luminous spaces' in the heavens are the effect of large numbers of stars; but on p. 65 he regarded them not as independent systems but as possible components of a compound system to which our stars belong. Wright had evidently not thought out the implications of this proposal. For example, if our Sun is a member of a spherical shell of stars which we suppose to form the innermost component of our compound system, then the sky to one side of the Milky Way (in the direction of the supernatural centre) ought to be empty of nebulae; if it is the outermost component then the nebulae visible to one side of the Milky Way would be other components, of which all except perhaps the innermost would necessarily be in the form of rings rather than shells of stars, while nebulae visible on the other side would be independent star systems.[55]

But granted there are other 'creations' like our own star system, what, one wonders, prevented Wright from carrying the hierarchy of moon-planet-star-creation one step further and, like Kant and Lambert, imagining these creations to be themselves components of a still higher system? Were the creations wholly in the natural order, this step would be an easy one. As it is, the question must be answered by counter-questions: What meaning could one possibly give to an hierarchy in the supernatural order? How can one supernatural centre be satellite to another? No, for Wright the way ahead is blocked by the very multiplicity of Abodes of the Blessed, and the next step in the elaboration of systems of systems had to be left to thinkers for whom the natural and the supernatural are less closely aligned.

NOTE ON 'A THEORY OF THE UNIVERSE'

Many of Wright's effects were purchased by Mr George Allan of Darlington, and when books and manuscripts from Allan's library were auctioned by Sotheby's on 18 December 1844, they included numerous papers of Wright, among them many if not all of the manuscripts now in Newcastle Central Library. These were purchased by Newcastle in 1898 from a Mr H. Gray for £3. *A Theory of the Universe* is cleanly written and without pagination, but every fourth page is headed with a new letter of the alphabet; here the page headed 'A' is indicated by 'A1' and the pages following by 'A2', 'A3' and 'A4', and so on. Interpolations by Wright, either in revision or at a later date, are placed within angle brackets ($<$, $>$), and editorial emendations within square brackets ([,]). Wright's capitals and punctuation, which are often ambiguous and seldom significant, have been freely altered. Grateful thanks are due to the City Librarian, Newcastle, for kind permission to publish this manuscript.

xxxiii

NOTES

1. Much of our knowledge of the first half of Wright's life comes from his own journal (British Museum ADD. MSS. 15627), published by Edward Hughes in 'The early journal of Thomas Wright of Durham'(*Annals of Science* **7** (1951), pp. 1–24). The quotations which follow are taken from this journal. Much bibliographical and other information about Wright was collected by the late F. A. Paneth and is conveniently assembled in *Chemistry and Beyond*, a selection of Paneth's writings edited by Herbert Dingle and G. R. Martin with assistance of Eva Paneth (New York, 1964). Paneth's account of Wright's theories, however, must be used with great caution.

2. See for example the list of William Whiston's publications at the end of his *Astronomical Principles of Religion, Natural and Reveal'd* (2nd edn, London, 1725).

3. Copies of the text are common, but the complete scheme is excessively rare. A facsimile of the text and scheme, with introduction by the present writer, was recently published (London, 1967).

4. For details of Wright's published (and some unpublished) writings, see *Chemistry and Beyond*, pp. 114–19. A list of Wright's manuscripts is being prepared by M. A. Hoskin and D. M. Knight.

5. *Op. cit.*, p. 2.

6. *Chemistry and Beyond*, p. 114.

7. According to the title page.

8. Copies survive in Durham University Library. I am grateful to the Librarian and to Dr A. I. Doyle for allowing me access to their rich collection of Wright material.

9. *Clavis Cælestis* (London, 1742), p. ix.

10. Ibid., p. 16.

11. Ibid., p. 75.

12. In the library of the Royal Astronomical Society. The Table of Contents differs somewhat from the published version and corresponds to an earlier organisation of the material. Letter IV as published appears to be a late addition, and Letter V to be an amalgamation of the original Letters IV and V; for in the prospectus Letter IV is 'Of the Order, Distance, and Multiplicity of the Stars, within our finite View', and Letter V is 'Of the *Via Lactea,* &c. proving the known sidereal Creation to be finite'. Most of the other changes are trivial, but in the published version 'amongst the Stars' has been added to the title of Letter VI and 'proving the sidereal Creation to be finite' to that of Letter VII.

13. Attention was first drawn to these documents and to their theological nature by Dr D. M. Knight of Durham University in 1965; see *Actes du XIᵉ Congrès International d'Histoire des Sciences* **3** (Warsaw, 1968), pp. 37–40.

In later life Wright returned once more to the problem of integrating the natural and the supernatural, in *Second or Singular Thoughts upon the Theory of the Universe*, a sequel to the *Original Theory* in the form of three further letters and other material. The manuscript of this work came to light only in 1966 when Mssrs Dawsons of Pall Mall invited the present writer to help sort the chaotic mass of Wright material they had purchased at auction. In these *Second Thoughts*, Wright abandons much of his earlier doctrine, and now teaches that the stars are celestial volcanos set in a heavenly shell which surrounds the Sun! This manuscript is now in the possession of Durham University; an edition by the present writer was recently published (London, 1968).

14. C. Huygens, *The Celestial Worlds Discover'd* (2nd English edn, London, 1722), p. 157.

15. Whiston, *Astronomical Principles*, p. 122.

16. Whiston, *Astronomical Lectures* (2nd English edn, London, 1728), p. 42.

17. W. Derham, *Astrotheology* (London, 1715), p. 55. Cf. Wright, *Original Theory*, p. 50: 'we may reasonably expect, that the *Via Lactea* . . . will prove at last the Whole to be together a vast and glorious regular Production of Beings . . . and that all its Irregularities are only such as naturally arise from our excentric View'; and ibid., p. 62: '. . . in like manner, as the Planets would, if viewed from the Sun, there may be one Place in the Universe to which their [the stars'] Order and primary Motions must appear regular and most beautiful'.

18. E. Halley, 'Considerations on the Change of the Latitudes of some of the principal fixt Stars' (*Phil. Trans.* **30** (1717–19), pp. 736–8).

19. Whiston, *Astronomical Principles*, pp. 88–9; cf. Whiston, 'A discourse . . . of the Mosaic history of the Creation', p. 37, in his *A New Theory of the Earth* (London, 1696); the question had of course been discussed by Richard Bentley and Isaac Newton in 1692 and 1693 (*The Correspondence of Isaac Newton*, edited by H. W. Turnbull, **3** (Cambridge, 1961), letters 398, 399, 403, 405 and 406), and the present writer is preparing an edition of related Newton manuscripts.

20. I am grateful to the Curators of Durham University Library for permission to reproduce Plates 1–3.

21. I am grateful to the Council of the Royal Astronomical Society for permission to reproduce Plate 4.

22. Whiston, *Astronomical Principles*, pp. 25, 131–2, 154 seqq.

23. T. Swindon, *An Enquiry into the Nature and Place of Hell* (London, 1714). For earlier ideas on the subject, and especially those of Milton whom Wright so often quotes, see C. A. Patrides, *Milton and the Christian Tradition* (London, 1966), pp. 280 seqq.

24. Swinden, *An Enquiry*, pp. 225 seqq.

25. Cf. Robert Fludd, *Utriusque Cosmi Maioris scilicet Minoris Metaphysica* (Oppenheim, 1617), pp. 131, 141. I first owed this suggestion to Professor Walter Pagel; as Fludd was little read at this period, the suggestion might have seemed implausible, but the subsequent discovery of Wright's *Second Thoughts* with its reference (Letter II, pp. 15–16; 1968 edition, p. 51) to Fludd's *Tomus Secondus de Microcosmi Historia* (Oppenheim, 1619) shows that at least in later life he knew something of Fludd's work.

26. Cf. Nils Thune, *The Behmenists and the Philadelphians* (Uppsala, 1948). Richard Roach, leader of the Philadelphians, was Rector of St Augustine's, Hackney, until his death in 1730, the year in which Wright worked in London for makers of mathematical instruments. For contemporary ideas of Hell see D. P. Walker, *The Decline of Hell* (London, 1964).

27. Jacob Behmen, *XL Questions Concerning the Soule* (London, 1647).

28. Boehme, *Alle Theosophische Schrifften* (Leipzig?, 1730), *Christosophia* plate 7.

28a. A survey of the development of Wright's cosmological thought as expressed in these three writings is given in Michael Hoskin, 'The Cosmology of Thomas Wright of Durham', *Journal for the History of Astronomy* **1** (1970), pp. 44–52.

29. J. Keill, *Introduction* (1st English edn, London, 1721), p. 17.

30. *Clavis Cælestis*, p. 75.

31. A copy of which was recently on sale at $1500.

32. The story is related by W. Hastie in the introduction to his translation of the work and other related pieces, *Kant's Cosmogony* (Glasgow, 1900); a reprint of this volume with introduction by Dr G. J. Whitrow is in preparation, and a partial reprint with an introduction by Professor Milton K. Munitz has recently appeared (Ann Arbor, Mich., 1969).

33. *Freye Urtheile*, Achtes Jahr (Hamburg, 1751), translated by Hastie, *op. cit.*, Appendix B.

34. P. L. M. de Maupertuis, *Discours sur les différentes figures des astres* (Paris, 1742); Hastie, *op. cit.*, pp. 32–33. For a critical assessment of the evidence, see Kenneth Glyn Jones, 'The observational basis for Kant's Cosmogony' (*Journal for the History of Astronomy*, **2** (1971), in press).

35. Hastie, *Kant's Cosmogony*, p. 190.

36. Ibid., p. lxx.

37. Reprinted in M. A. Hoskin, *William Herschel and the Construction of the Heavens* (London, 1963).

38. Ibid., pp. 115–16.

39. Ibid., p. 54.

40. Ibid., pp. 32, 42. But Wright does in fact later assume that a star of the ninth magnitude is nine times further than one of the first magnitude (*Original Theory*, p. 73).

41. *The Universe and the Stars*, being an original theory on the visible creation, founded on the laws of nature, by Thomas Wright, with notes by C. S. Rafinesque (Philadelphia, 1837). In the Dedication to the American Public, Wright is seen as the reviver of the ancient learning of Egypt, and the republication of his book was the first activity of the American Institute of Learning, newly founded 'for the purpose of restoring, spreading and increasing knowledge of all kinds and on all topics. . . . To the members of it will be due the foundation of an Eleutherium of Knowledge or free school of useful knowledge, an extensive mutual library and museum, and the gradual publication of valuable works on all branches of human knowledge' (Preface, p. 5). Rafinesque, who describes himself as Professor of historical and natural sciences, compares Wright to Plato, Copernicus, Newton and Herschel (p. 7); he has some inkling of Wright's conception of our star system, but is dazzled by the brilliance of Wright's religious insight. He apologises for the absence of the plates (p. 8), but promises their separate publication in the future.

42. A. de Morgan, 'An Account of the Speculations of Thomas Wright of Durham' (*Philosophical Magazine* (3) **32** (1848), pp. 241–52), reprinted in Hastie, *Kant's Cosmogony*, Appendix C.

43. *Kant's Cosmogony*, p. 197.

44. Ibid., p. 203.

45. To this we must record one honourable exception, Vera Gushee, whose article 'Thomas Wright of Durham, Astronomer', appeared posthumously in *Isis* in 1941 (**33**, pp. 197–218). The portrait she reproduces is probably of another Thomas Wright, and many other criticisms can be made of her work, but she does realise that Wright taught no disk-theory of the Galaxy. Yet Paneth (*op. cit.*, p. 118) says 'This article fails to understand Wright's *Original Theory*'!

46. Edward Sherburne, *The Sphere of Marcus Manilius* (London, 1675).

47. *Kant's Cosmogony*, pp. 182, 194.

48. Swinden, *An Enquiry* (2nd edn, London, 1727), p. 355; Whiston, *Astronomical Principles*, p. 19; Wright, *Clavis Cælestis*, pp. 16, 17, 33. Wright discusses the matter further in *Second Thoughts*, Letter I, p. 34 and Letter II, p. 14 (1968 edition, pp. 42, 50).

49. J. Michell, 'An Inquiry into the probable Parallax, and Magnitude of the fixed Stars, from the Quantity of Light which they afford us, and the particular Circumstances of their Situation' (*Phil. Trans.* **57** (1767), pp. 234–64).

50. W. Herschel, 'Account of the changes that have happened during the

last Twenty-five Years, in the relative Situation of Double Stars' (*Phil. Trans.* **93** (1803), pp. 339–82).

51. P. 41.

52. This is brought out especially by Plate 1 supplied for this edition.

53. *Original Theory*, p. 62.

54. Ibid., p. 79.

55. A hint of these complications is to be found in the negligible criticisms made in an open letter to Wright and signed 'S. L.', *The Gentleman's Magazine* **21** (1751), pp. 315–17.

THE ELEMENTS OF EXISTENCE
OR
A THEORY OF THE UNIVERSE

[Wright MSS Vol. VII, Central Library, Newcastle upon Tyne]

A^N hypothesis to solve, nearest to the level of ye human under-standing, the Being of a God and ye origin of Nature grounded upon visible effects and final causes. Demonstrating not only ye visible creation to <be only> a finite view <of ye general frame of Nature>, but also the general disposition of the whole <is evenced from it.> Together with ye several mansions of hapiness & misary in a future state | and to be rationaly expected after death. All which is represented in a section of ye Univers twelve feet radius, extending from the Imperial Seat or *Sedes Beatorum* to ye verge of chaos bordering upon ye infinite abiss. Comprehending first ye Paradise of imortal spirits in there several Degrees of Glory surounding the Sacred Throne of Omnipotence. | Secondly, the Gulfe of Time or Region of Mortality, in which all sensible beings such as ye planetary bodies are imagind to circumvolve in all maner of direction round the Devine Presence, or ye Eternale Eye of Providence. Thirdly the shades of Darkness & Dispare supposd to be the Desolate Regions of ye Damn^d. |

<Wrote in ye year 1734: the author being then 22 years old.> |

[2]

[3]

[4]

[1]

THE EXPLANATION

At the top is a folded snake; the emblem of Eternity, supported by four boys representing the perpetual youth or mysarable [?] state of ye four elements, wch are supos'd to be the active powers of creation and a little below these upon a double label is this general title:

THE THEORY OF EXISTENCE.

3

On each side of these by two & two are plac^d four of ye seven arch angels, namely

Raphael *Ithuriel* *Iophel* and *Uriel*.

These are represented as ye active agents of God in the image of his Devine Being, and personal both in majesty & power as many corporial Regents. | A little below these on one side, is ye following motto.

[2]

The Eyes of ye Lord are in every place,
Beholding ye eval and the good

Prov: XV, 3.

and on ye other side opposite to it; this

Be wise now therefore O ye Kings,
Be learned ye that are Judges of ye Earth.

Psalms: II, 10.

In ye middle betwixt ye above labels is the Seraph *Abdiel* supporting <a Book of> the Holy Scriptures the leaves of which are illuminated by two Suns representing the two great lights of natural knowledge <as flowing> from reveal'd relegion; the Law and the Gospel, in this is written

!*That I may publish with ye voice of thanksgiving*
and tell of all his wondrous works. |

[3] Immediately below this is an historical and emblematical view of the primitive creation as related by *Moses* with ye expulsion of the evall geneouses or diabolical spirits from y^e *Emprimum* or primogenial Seat of Good. Here are severall emblematic^l figures to express the sacred and devine Vertues of that Eternall Trinity in ye union of which we comprehend the Deaty. The *Simbol* of this Holy Trinity, chiefly consists in a displaceation of y^e †Tetrahedron, by which is ment the mistical union of

* Those impressions are all from that [edge?] 4 inches.

† The property of every equilateral triangle is such that ye nearest approach, contact, or union of any three, adequate to each other make one more of the same kind & similar to ye whole.

[4] [The] power & propriaty of the figure | is evedent from the expression of one in three & three in one.

4

those necessary and self existing Principals which together we call the *Elohime*: over this is represented

The Eye of Providence. |

[4] How this will agree with the Trinity of yᵉ Church which you explain in a very different maner, I shall not make it my business here to determine but leave the true senses of it, to be expounded by abler heads.
I have indede indevoured to define it thus

one omnicient ; one omnipotent, and one omnipresent Being, unighting as by necessity *constitude one God <whose state is> infinite & eternall.*

For to have cognisense without power & power without dominion, implyes an universall negative. & again, dominion without power and power without knowledge is still more absurd; and consequent of much more fatal errors, in the idea of an active ignorance. But in the union of yᵉ three we comprehend the perfection of all beings, namely the *Living God.*|

[A1]   Now as the antient Egyptians chose to express by characters; significant words, and simbols, the misteries of their relegion; and particularly such conceptions <of yᵉ Deaty> as all the force of language could not <otherwise> express, I have so far follow^d their examples as to fix the following *Hyroglyphic* in place of yᵉ Devine presence full in the middle of yᵉ principal scheme.

What I mean by this is the fountain of all life, joy and love, from whence we derive our being, and all the blessings we enjoy, thrô our existence. & by yᵉ same infinite Wisdom and all mighty Power every where existing, I would willingly understand this part of *St Athenatious Creed*: God of God, Light of Light, very God of very God, before all worlds, incomprehensible. |

[A2]   Near to the Emblematic Trigon, upon the right, is a figure of Piety or Devotion, representing Relegion, and on the left the Geneous of Natural

Philosophy, as presidinge over knowledge, and perticular ye science of astronomy, the former attentively listening to ye latter who is suppos^d to be explaining the disposition & laws of the New Creation. Below these at some distance is shown the revolt or expulsion of Satan & his angels, who are all together represented falling from the Imperiale Heaven of Heavens, Satan alone preserving an erect *atitude*, as resisting the Devine wrath, with this false motto upon his sheald.

——*There is no God.*

Amongst the falling angels, I have introduced in a prostrate posture of dread & horor Original Sinn, and over her head as descending the Chirub Zophil, one of the host of Heaven, bearing a masy chain to keep those wicked spirits in eternal bondage. Other angels on each side are sound[ing] thro clarion trumpets the everlasting fame of the victorious triumph. |

[A3] Under the falling angels is this motto:

Hell and Distruction is before ye Lord.

Above these diabolical spirits in ye radiating sillia proceeding from the Deaty is a figure of *Urainia* the heavenly muse, seated in a shining cloud and suported by representatives of Trouth & Justice each bearing the propper emblems of Sacred Wisdom and Power. Here are also respectively display'd the mistic Bokes of Fate, containing the decrees and ye perpetual laws of ye Eternal Will.

The Mottos are

Alpha et omeg.

and

He standeth in the Congre[g]ation of ye Mighty
And judgeth amongst the Gods.

Psalms LXXXII 1st. |

[A4] Immediately below this historical view, commences the grand section or rather sector of ye creation, it being a partial radii, passing thrô all the Regions of Philisity, of Probation & Punishment, from the very verge of

* The fool hath said in his heart, there is no God.

the orbicular creation to the Relms of perfect Bliss. And here that part of it peculiar to yᵉ more imediate Presence of God i.e. the visible Deaty, is fully display'd and suported by 6 Gardian Angels, at the top is a figure of Fame sounding the praises of yᵉ Eternale Godhead upon his act of creation.

This end of the scheme comprehending the center, the other as being the extremity of the orb, and supposed to be the most remote from the focus of power, is void of all created beings & is suppor[t]ed by evale geneouses. Here as upon the verge of Chaos surrounded by discord and confusion is an emblem of extreme disgrace. | This most horrid spectacle is express'd by a human figure upon whose head the Devine vengence is represented falling & breaking like thunder & lightning, and on each side are two demons supposd to be tormentors.

Thus I have given you a general discription of the signifier or section, as the whole is represented & supported 16 feet long, and shall now proceed to a more particul[ar] one.

First, you are to imagin that this particul[ar] view of the Universe, is a section of the visible creation, formed upon a segment of yᵉ infinite plain passing thrô the center of yᵉ *Cedes Beatorum* & coinciding with that of our Earth or yᵉ solar orbit. | The lucid body, or radient point of this scheme is supposed to be one of those Eternal Mansions appropriated by yᵉ Devine Being for his first born, or Eldest Sonˢ, begotten of his Sacred Will before all Worlds: Deo of Deum. In these imediate or visible Residences of the Deaty from whence he becomes sensible to his cretures and omnipresent thrô all infinity, Might, Majesty & Dominion is made manifest by an en[d]less Sphere of Glory. The center of which being imagined the *Primogenit[al] Fountain* of all perfection from whence the ever active powers of Creation & generation receiv'd there secret Laws I suppose to be the true *Sanctum Sanctorum* of that inatiquate Being of Beings which yᵉ silent soul of every mortal dreads, which Providence makes manifest in ex[is]tence, but what yᵉ human mind is [un]able to conceive. | Here the Eternall Spirit <comprehensive of all good> was pleasd to promulgate his powerful will by promising of one Sacred all active & inliving Word, productive of all temporal beings, i.e. the generale *Fiat*.

In place of Whome you have this combination

Thus in conformity with yᵉ Trinity of Etern[ity], the Spirit or Fountain of infinite Light concerning such an effect of thought as from the result of reason, namely the will, this determation of infinite Wisdom imediatly proceeds to generate the primary Cause and lastly by the Word comprehending allmighty Power, the joyous act of creation was made manifest in an en[d]less chain of varigated [?] beings. |

[B4]   Hence by a joint adherence of the Spirit to the Will and yᵉ Will to yᵉ Word all firmly unighting in yᵉ act or de[e]d, yᵉ universal laws for a perpetual change are thus promulgated from their Heavenly Origin.

The general section you are to understand is principally devided into three parts which for distinction let me call the internal, yᵉ external & yᵉ middle regions, and in order that we may proceed in yᵉ most regular and natural order, 'twill be necessary to begin our explanation at yᵉ place of our present habitation (viz) the Earth. |

[C1]   The system of the Sun round whome yᵉ Earth performs her annual motion is situated near yᵉ center of ye middle region; this for better distinction I would call yᵉ Sphere, or Gulf of Mortality which within its orbicuity comprehends all present beings in their natural state of probation. This stary firmament or comon frame of terrestrial worlds I supose to be bounded by the two extremes of Nature, namely yᵉ desolate Regions of Darkness and yᵉ joy full Relms of Light. The former or external region is supos'd to limit yᵉ human state or creation on yᵉ side of Chaos, this circumscribes the whole & borders upon the *old Eternal Night.* The latter or internal region is by Creation circumscribed as the common center of gravitation to all the radiant bodies round it, and this I can easily conceive to be yᵉ true imperial seat |

[C2]   or *Sedes Beatorum,* common to all superior spirits in the most exalted state of happiness.

Betwixt these two yᵉ Gulph of Human Life is fix'd and suppos'd to be so situated betwixt yᵉ Good & Eval Agents, that neither vice nor vertue can originally produce predomenant passions, but yet externally attended

8

with every faculty of sense that may possibly corrupt the primitive inclinations, or add fresh lustre to yᵉ principals of virtue, by exciting the soul or new moduling the mind, according to yᵉ dictates of Devine Reason.

<This tho a state of perfect inocense & freedom is> equally open on all sides both to yᵉ influence of Heaven and yᵉ insults of Hell, obnoctious and exposed for a trial of vertue, originaly designd for human good and finaly the glory of God. |

[C3] These several regions are again subdevided into various degrees or gradations of hapiness & misery according as they are situated nearer or farther remote from the extreems of either, but so sepparated by the Gulph of Mortality, that yᵉ nature and laws to which either are subject are intirely unknown to yᵉ other. The projection or plain of the picture you are to suppose parallel to yᵉ *Orbis Magnus* and the perspective view penetrates so far into yᵉ opposd concavity as to discover meriads of systems in all maner of dispostions with an infinite number of worlds [which] are variously distributed round their several Suns or Centers. And upon all sides the principal stars of the visible creation are exhibited in ther natural order as seen from yᵉ Earth by yᵉ naked eye. Those of yᵉ first magnitude nearest to our own system, and the rest proportionable

[C4] removed according to their resp[ec]tive phenomena. | Beyond these are others more remote crownd by a penumbral shaddow such as we call telescopic stars and again without them more, suposᵈ to be at immense distance & by no means perceptible to yᵉ human eye. At a certain distance from yᵉ Sun equal to a vissual ray of yᵉ smallest visible star is a faint circle of light terminating the utmost extent of yᵉ visible creation, in a finite view from yᵉ Earth; and all within this sphere is more or less inlightenᵈ by yᵉ rays of our Sun according to their distance from him.

This vast creation of worlds I call the Gulph of Time or Region of Mortality, and next to this penetrating towards the abodes of the Blest you arive upon the Borders of Light, but if receeding in yᵉ opposite direction you are led to Shades of Darkness & Desolation adjoining to yᵉ presincts of the infinite abiss. |

[D1] The first mansion of immortality is imaginᵈ to be [the] place where yᵉ Gardian Angels of yᵉ human soul, present their charge to yᵉ respective pations [?] of their faith. Here yᵉ souls of yᵉ Proffits, Patriarchs and

Primitive Fathers are first iniciated to yᵉ Devine inspection and as subjects of Heaven obey yᵉ sumons.

In this region as being yᵉ first dawning sphere of immortal glory, the Royal Psalmest is represented playing upon his harp, praising God and his great creation; behind him is a gardian angell holding a wreath of palm over his head and an emblem of peace & endless joy; befor him is a lesser angel suporting his book of pious *Orisons*. | A little above upon one side is an angel resting upon a cloud, & supporting an orbicular simbol, representing yᵉ eternal vicessitude or perpetual change of all created beings, and on yᵉ other side is a long light vista leading to yᵉ Primogenial Center of the Sedes Beatorum. At yᵉ extremity of this luminous radia is yᵉ figure of an eagle representing Contemplation, bearing in its claws a long scrawl or label with this inscription:

[D2]

Strate is yᵉ gate
and narrow is yᵉ way,
that leadeth unto Life. |

[D3] On each side of yᵉ great section supposd to be stationd in yᵉ terrestrial regions are divers figures representing philosophers both ancient & modern, holding in their hands various schemes, demonstrating the several laws & principals of their respective relms explanatory of their motions, distance & phenomena. Here yᵉ Geneous of Mathematicks is represented with a ballance, holding yᵉ Sun & several planets in equilibrio and below is this motto:

He hangeth yᵉ Earth upon nothing.

Opposite to this is yᵉ Geneous of Geometry pointing to a diagram or construction, shewing the supernatural censure, catastrophy & conflagration of a world distroy'd. |

[D4] But now to reascend above yᵉ Royal Psalmist our next step is into a more refined region and of still more exalted hapiness, in this sphere are suppos'd to be an infinite number of saints and angels, some of which are here represented in various postures of addoration; to whome a crown of glory is desending supported by two angelic boys each bearing the emblems of peacefull innocense along with yᵉ regal rewards of vertue.

10

The next orb of Heaven approaches still nearer to perfection and is a relm of more & more refind beatitud. This I imagin to be a place of perfect joy & tranquili[t]y, full of the everlasting glories of yᵉ Deaty, in that imortal state, which is next yᵉ supreeme Heaven, all mutuale delite

[E1] of consumate happiness & universal peace. | Here as upon his first admitence to the Devine Radience, is represented a transpo[r]ted Patriarch, struck with admiration and astonishment upon so visible an approximation to yᵉ Devine Presence. He is neeling upon his knee unable to support himself under yᵉ strong radient glories and superlative luster of the place. Here resting upon a cloud is an angel leaning upon a golden simbol, denoteing harmony & union. He is pointing towards yᵉ Emperial Throne and over his head upon a flowing label is this sentence:

> Glorious things are spoke of thee
> *O City of God.* |

[E2] At some distance from this on yᵉ opposite side seated upon an other cloud is another ang[el] represented as a scribe, he has a large book liing open before him in which is writ <this part of> yᵉ sacred Te Deum or Bened:

> We praise thee O God
> We acknowledge thee to be the Lord.

Here, many angells with divers instruments of music are performing a celestial anthem to yᵉ addoration of their devine presedents.

Within this circumambient region at various distance[s] from yᵉ Center of Glory are many and vast groups of saints and angels, besides miriad

[E3] other flying figures bearing various simbols of | peace, plenty, joy & love.

Allmost centering with perfection are infinite numbers of still more refind beings nearly uniteing with yᵉ Deaty but of no determind image, seaming yᵉ lest in glory.

The next and last step is to yᵉ fountain of all perfection; the Center of Life & Immortality, but here yᵉ imagination must conceive, what human language can't express, a place so full of the Divine Lust[re] that

[E4] probably angels of yᵉ 1ᵗ order, may only behold it with a coverᵈ eye. | The Throne of Grace or of Devine Majesty, is expresst by yᵉ following hyroglyptic:

11

Right over this by which you are to understand the Emperial Seat, but at an angule distance from it, is a Book of the Holy Communion of Saints and Angels supported by two arch seraphims and in this to express [F1] in some measure the immeasurable concourse of the Heavenly | Host, with yᵉ general joy and universal aclamations of angels upon the act of creation, is this verse:

> The Morning Stars sung together
> and all the Sons of God
> Shouted for joy.

Having now carried you thro' yᵉ celestial regions I must now begin to receed and direct your attention towards yᵉ infernal ones, but I am afrade I shall not be able to make the discription of them answer yᵉ comon ideas. First I shall beg leave to suppose this rationaly, that if happiness increase as yᵉ distance from perfection decreases, misery may also increase & torment as yᵉ obje[c]ts are removed from the prime Presence of yᵉ [F2] Deaty. | Consequently it may prove to be more than a conjecture, that those external regions beyond yᵉ verge of this creation, may be yᵉ commensment of an infinite abiss of desolation bordering either upon Chaos, or yᵉ interstices of general night; this I would suppose the place alloted for temporal punishments in a future state. In this dark region far remote from God, and void of all created being, Sattin and his angels are supposed to range in desolation, this being imagind the Horrid Mansion to which they were driven after their expulsion from yᵉ Emperial Heaven. The inferior part of these eval agents are represented like snakes inter- [F3] woven with each other and grovelling in the utmost confusion. | Now as I don't know how to imagin that a spirit falling from a finite state thô in yᵉ greatest excess of sinn can merit an equall degree of pain, with one fallen from a state of perfection, I have subdevided this Relm of Darkness into three different states, and that bordering upon the Eternal Night I have given to the most demonical spirits, the primary leaders or inspirators of the primitive rebelion such as the Devil and his original revol[t?]ing Agents. Here yᵉ great arch angel Michael, in all his radient glories, is represented brandishing the thunder of God upon yᵉ arch offende[r] |

[F4]   which appears of a gloomy aspect under his feet, attended by Envy in a like disabled posture teasing her snakes. Upon Michaels sheeld is inscribed this word

Invinsible.

Bordering upon this region towards the Gulph of Time is a much milder sphere where you are to imagin the secondary angels or procelites who fell without premeditated designe, more by yᵉ ill advice and influence of others than self choruptors. This region is faintly inlightened by a dubious

[G1]   dawn proceding from a confused | light reflected from the luminous bodies in yᵉ Region of Mortality or Relms of Death. Here Sin & Death are represented preparing an invasion against yᵉ Human Species and at some distance is a weeping boy struck with a sense of future ills and is lamenting yᵉ state of Man, being subject to mortality & obnoxions on all sides [and] to yᵉ insults of Hell. By him is this line:

Wo to yᵉ inhabitants on Earth. |

[G2]   Next above these joining to yᵉ verge of creation is yᵉ state appointed for yᵉ human souls or yᵉ punishment of mortals. Here the eye may be supposed able to penetrate into yᵉ abiss of motion, as in some part of the abiss it self the human eye <in> its mortal state may perhaps se[e] some of yᵉ eternal beams proceeding from yᵉ Seat of Glory. In this case yᵉ wicked thô in a state of pain may possibly behold on one side a bright and glorious creation, and on yᵉ other nothing but gloomy darkness & horrid obscurity.

Here as a seene of punishment is represented a human figure chained upon a rock of cediment [?] to a penal bed of undisolving icce, & near him is written this: |

This is the state of a wicked man.

[G3] We may suppose that after our mortal change the souls of [the] wicked do not imediately enter upon bodily paine or any kind of material

13

punishment, but as in a state of oblivion are prepar'd by painfull visions or tormenting dreames arising by a kind of internal[?] reflection from yᵉ near approaching scene of Wo in order to <strengthen yᵉ mind &> to enable them to support such dreaded pains. In like manner we may imagine the <souls of yᵉ> good return to rest after death and find an Elizium in the grave and anticipate yᵉ joys to come. |

[G4]   The situation of these several regions of joy & pain in a very lively manner p[rese]nts to our imagination that misterious parable of *Dives* & *Lazarus* where yᵉ Patron of Christianity endeavouring to give the Jews and Gentiles, not only an idea of yᵉ weakness of human faith in this life but also an imaginary view of yᵉ fatal distance of Heaven & Hell in a future life, tells them by way [of] allegory in yᵉ person of Abram that those two places are so situated and at so vast a distance from each other

[H1]   that betwixt them ly an impassable gulph impenetrable on | either side, but by yᵉ souls of yᵉ dead who by a vertuous or impious death are permitted to pass from this world to happiness or missery in the next. This image of futurity I have emblematically represented consentanious to Scripture by yᵉ three different states, viz: the present problematical, yᵉ future good & yᵉ future ill.

At yᵉ lower end of yᵉ scheme are placᵈ two erect figures as spectators of yᵉ whole creation and above one of them is this verse:

> The visible things of God, from the
> creation of yᵉ World is clearly seen
> being understood by yᵉ things that
> are made even his eternal power &
> godhead.
>
>   Rom 1 : 20ᵗʰ. |

[H2]   and over yᵉ other this.

> The Heavens declare yᵉ glory of God
> and the firmament sheweth his
> handy work.
>
>   Psalms XIX 1.

In the middle betwixt yᵉ fore mentiond figures is yᵉ arch angel Gabriel

14

treading upon a clood and pointing to ye lower parts of ye scheme and over his head are these laws.

Stand still and consider the wondrous works of God.

Job: XXXVII, 14.

He is far from ye wicked.

Prov. XV. 29.

Behold what desolation he has made.

Lastly ye scheme closes with two figures representing Chaos and Old Night. |

[H3] <This juvenil performance was the produce of the authors imagination before he had reap'd any advantages either from reading or study, but prov'd afterwards the foundation of his Theory of ye Univers a much more perfect work.>

AN

ORIGINAL THEORY

OR

NEW HYPOTHESIS

OF THE

UNIVERSE,

Founded upon the

LAWS of NATURE,

AND SOLVING BY

MATHEMATICAL PRINCIPLES

THE

General PHÆNOMENA of the VISIBLE CREATION;

AND PARTICULARLY

The VIA LACTEA.

Compris'd in Nine Familiar LETTERS from the AUTHOR to his FRIEND.
And Illuftrated with upwards of Thirty Graven and Mezzotinto Plates,
By the Beft MASTERS.

By THOMAS WRIGHT, of DURHAM.

One *Sun by Day, by Night* ten Thouſand *ſhine,*
And light us deep into the DEITY.   Dr. YOUNG.

LONDON:
Printed for the AUTHOR, and ſold by H. CHAPELLE, in *Groſvenor-Street.*
MDCCL.

THE

PREFACE.

HE Author of the following Letters having been flattered into a Belief, that they may probably prove of some Use, or at least Amusement to the World, he has ventured to give them, at the Request of his Friends, to the Publick. His chief Design will be found an Attempt towards solving the Phænomena of the *Via Lactea*, and in consequence of that Solution, the framing of a regular and rational Theory of the known Universe, before unattempted by any. But he is very sensible how difficult a Task it is to advance any new Doctrine with Success, those who have hitherto attempted to propagate astronomical Discoveries in all Ages, have been but ill rewarded for their Labours, tho' finally they have proved of the greatest Benefit and Advantage to Mankind. This ungrateful Lesson we learn from the Fate of those ingenious Men, who, in ignorant Times, have unjustly suffered for their superior Knowledge and Discoveries ; they who first conceived the Earth a Ball, were treated only with Contempt for their idle and ridiculous Supposition, as it was called ; and he who first attempted to explain the *Antipodes*, lost his Life by it ; but in this Age Philosophers have nothing to fear of this sort, the great Disadvantages attending Authors now, are of a widely different

A 2

Nature,

Nature, rifing from the infinite Number of Pretenders to Knowledge in this Science, and much is to be apprehended from improper Judges, tho' from real ones nothing; for nothing is more certain than this, as much as any Subject exceeds the common Capacity of Readers, fo much will the Work in general be condemned; the Air of Knowledge is at leaft in finding Fault, and this vain Pretence generally leads People, who have no real Foundation for their Judgment to argue from, to ridicule what they are too fenfible they do not underftand. Thus the fame Difadvantages too often attend both in publick and private an exceeding good Production equally the fame as a very bad one: But the Author is not vain enough to think this Work without Faults, has rather Reafon to fear, from the Weaknefs of his own Capacity, that there may be many; but he hopes the Defign of the Whole will, in fome meafure, plead for the Imperfection of the Parts, if the Merits of the Plan fhould be found infufficient for his full Pardon, in attempting fo extenfive a Subject.

In a Syftem thus naturally tending to propagate the Principles of Virtue, and vindicate the Laws of Providence, we may indeed fay too little, but cannot furely fay too much; and to make any further Apology for a Work of fuch Nature, where the Glory of the Divine Being of courfe muft be the principal Object in View, would be too like rendering Virtue accountable to Vice for any Author to expect to benefit by fuch Excufe. The Motive which induces us to the Attempt of any Performance, where no good Reafon can be fuppofed to be given for the Omiffion, or Neglect of it, will always be judged an unneceffary Promulgation, and confequently every Attempt towards the Difcovery of Truth, the Enlargement of our Minds, and the Improvement of our Underftandings will naturally become a Duty. If therefore this Undertaking falls fhort of being inftrumental towards the advancing the Adoration of the Divine Being in his infinite Creation of higher Works, and proves unable to anfwer all Objections that may poffibly arife againft it, yet will its Imperfections appear of fuch a Nature to every candid Reader, as to afford the Author a fufficient Apology for producing them to the World: And it is to be hoped farther, that where a Work is entirely upon a new Plan, and the Beginning, as it were, of a new Science, before unattempted in any Language, the Author having dug all his Ideas from the Mines of Nature, is furely intitled to every kind of Indulgence.

To

To thofe who are weak enough to think that fuch Enquiries as thefe are over-curious, vain, and prefumptive, and would willingly, fuitable to their own Ignorance and Comprehenfion, fet Bounds to other People's Labours, I anfwer with Mr. *Huygens*, " That if our Forefathers had " been at this Rate fcrupulous, we might have been ignorant ftill of the " Magnitude and Figure of the Earth; or that there was fuch a Place as " *America*. We fhould not have known that the Moon is enlightened by " the Sun's Rays, nor what the Caufes of the Eclipfes of each of them " are; nor a Multitude of other Things brought to Light by the late " Difcoveries in Aftronomy; for what can a Man imagine more abftrufe, " or lefs likely to be known, than what is now as clear as the Sun."

> Had we ftill paid that Homage to a Name,
> Which only God and Nature juftly claim;
> The weftern Seas had been our utmoft Bound,
> Where Poets ftill might dream the Sun was drown'd;
> And all the Stars that fhine in Southern Skies,
> Had been admir'd by none but favage Eyes.
>
> DRYDEN.

Befides the Noblenefs and Pleafure of thefe Studies, *Wifdom* and *Morality* are naturally advanced, and much benefited by them, and even Religion itfelf receives a double Luftre, " to the Confufion of thofe who " would have the Earth, and all Things formed by the fhuffling Concourfe " of Atoms, or to be without Beginning." In Aftronomy, as well as in natural Philofophy, though we cannot pofitively affirm every thing we fay to be Facts and Truth, yet in fo noble and fublime a Study as that of *Nature*, it is glorious, as Mr. *Huygens* fays, even to arrive at Probability.

Notwithftanding then the Difadvantages which ever have attended all new Difcoveries, either thro' the Ignorance of the Age, or the univerfal Paffion of Ridicule in fuch contented Creatures, as can't comprehend, yet ever attacking with a fool-hardy Refolution, the advancing Enfigns of Knowledge, if Ignorance was Virtue, and Wifdom Vice; I fay, regardlefs of this noify Shore, it is fure our Duty to fpring forward, and explore the fecret Depths of Infinity, and the wonderful hidden Truths of this vaft Ocean of Beings. But how the heavenly Bodies were made, when they were

made

21

made, and what they are made of, and many other Things relating to their Entity, Nature, and Utility, seems in our present State not to be within the Reach of human Philosophy; but then that they do exist, have final Causes, and were ordained for some wise End, is evident beyond a Doubt, and in this Light most worthy of our Contemplation.

> He who thro' vast Immensity can pierce,
> See Worlds on Worlds compose one Universe,
> Observe how System into System runs,
> What other Planets, and what other Suns;
> What varied Being peoples ev'ry Star;
> May tell why Heav'n made all Things as they are.
>
> POPE.

To expect that so new an Hypothesis should meet with universal Approbation, would be an unpardonable Vanity; nor is it reasonable every Reader should think the Author obliged to remove all his Prejudices and Partialities, so far as to give him the perfect Picture of the Universe he likes best. In many Cases it would be so far from being better for the World, if all Men judged and thought alike, that Providence seems rather to have guarded against it as an Evil, than any how to have promoted it as a general Good: But the following Theory regards the Whole rather than Individuals: And the many worthy Authors cited in the Work, who have all greatly favoured this extensive Way of Thinking, will, I hope, be a sufficient Excuse for forming these obvious Conjectures into a Theory, especially where so great a Problem is attempted as the Solution of the *Via Lacteal* Phænomenon, which has hitherto been looked upon as an insurmountable Difficulty. How the Author has succeeded in this Point, is a Question of no great Consequence; he has certainly done his best; another, no Doubt, will do better, and a third perhaps, by some more rational Hypothesis, may perfect this Theory, and reduce the Whole to infallible Demonstration: The first System of the solar Planets was far from a true one, but it led the Way to Perfection, and the last we can never too much admire. It is well known, that the first System of the Planets was also but a Conjecture, yet none will deny that it was an happy one.

The

The Difcovery of the Magnet Poles; the Government of the Tides; proportional Diftance and Periods of the Planets, &c. have all their Ufes, and undoubtedly were defigned to be known. Ignorance is the Difgrace of Mankind, and finks human Nature almoft to that of Reptiles. Knowledge is its Glory and the diftinguifhing Characteriftic of rational Creatures.

To Enquiries of this fort, then fure we may fay with *Milton*, That

GOD'S OWN EAR LISTENS DELIGHTED.

The Subject is, no Doubt, the nobleft in Nature, and as fuch, will always merit the Attention of the thinking Part of Mankind. Men of Learning and Science, in all Ages, have ever made it their peculiar Study. Towards the latter End of the Republic, and afterwards in the more peaceable Times of *Trajan* and the *Plinys*, we have no Reafon to doubt but that Aftronomy was in the higheft Reputation: And notwithftanding *Greece* had been the chief Seat of the Philofophers, yet may we fuppofe *Rome* in thofe Days little inferior in the Knowledge of the Stars, when we find Men * of the firft Figure in Life become Authors upon the Subject.

We have many Inftances to fhew, that Aftronomy was in the greateft Repute amongft the Antients of all Ranks, and almoft every where looked upon as one of the greateft, if not as one of the firft Qualifications of their beft Men. As a Confirmation of which, we find in the hiftorical Accounts of the *Argives*, a very warm Conteft betwixt the two Sons of *Pelops* 1205 Years before *Chrift*, thus teftified by *Lucian*: When the *Argives*, by publick Confent, had decreed that the Kingdom fhould fall to him of the two, who fhould manifeft himfelf the moft learned in the Knowledge of the Stars, *Thyeftes* thereupon is faid to have made known to them, the Conftellation, or Sign of the *Zodiack* call'd *Aries*: But *Atreus* at the fame time difcovering to them the Courfe of the Sun, with his various Rifing and Setting, demonftrating his Motion to be * contrary to that of the Heavens, or diurnal Motion of the Stars, was thereupon elected King.

* *Cicero* tranflated the Phænomena of *Aratus* into *Latin* Verfe. *Julius Cæfar*, as *Pliny* relates, wrote of Aftronomy in *Greek*, and is faid to have left feveral Books of the Motion of the Stars behind him, derived from the Doctrine of the *Egyptians*. *Ant. Chrif.* 45. He with *Sofigenes* reformed the *Roman* Year, which was firft invented by *Numa Pompilius*. *Germanicus Cæfar* alfo tranflated *Aratus*'s Phænomena into *Latin* Verfe *Anno Dom.* 15. *Tiberius* and *Hadrian* are alfo faid to have wrote on Aftronomy.

* Hence arofe the Fable of the Sun's going backwards in the Days of *Atreus*, as if ftruck with Abhorrence of his bloody Banquet. *Vide Ovid*'s Metamorphofis.

To

To recite more of the moſt eminent Patrons and Profeſſors of this kind of Learning here, will carry me too far from my preſent Purpoſe; for farther Information therefore, I ſhall refer the inquiſitive Reader, to that curious Catalogue in *Sherburn*'s Sphere of *Manilius*, where ſo many ruling † Men of all Ages and Nations ſwell, and illuſtrate the Number.

In a Word, when we look upon the Univerſe as a vaſt Infinity of Worlds, acted upon by an eternal Agent, and crouded full of Beings, all tending through their various States to a final Perfection, and reflect upon the many illuſtrious Perſonages, who have, from time to time, thought it a kind of Duty to become Obſervers, and conſequently Admirers of this ſtupendious Sphere of primary Bodies, and diligent Enquirers into the general Laws and Principles of Nature, who can avoid being filled with a kind of enthuſiaſtic Ambition, to be acknowledged one of the Number, who, as it were, by thus adding his Atom to the Whole, humbly endeavours to contribute towards the due Adoration of its great and divine Author.

I judge it will be quite unneceſſary to ſay any thing about the Order of the Work, ſince that would be only a Repetition of the Table of Contents, to which the Reader is referred, as to the propereſt Account that can here be given.

---

† Seven Emperors, nine Kings, and as many ſovereign Princes. *Charlemagne* wrote *Ephemerides*, and named the Months and Winds in *High Dutch*, 770. *Rich.* II. &c.

T H E

THE
CONTENTS.

a　　　　　　　DIREC-

DIRECTIONS for placing the PLATES.

Some of the Principal ERRATA.

Page	Line	the Words	Read.
2	ult.	to ceafe relating	ceafing to relate
4	3	Phænomenon	Phænomena
16	15	incomfible	incomprehenfible
21	12	comprehend	comprehending
33	28	compared	is compared
34	37	form	from
43	20	volving	revolving
49	24	immoveable	moveable
61	19	much	much as
62	28	XXIII.	XXI.
65	4	where	any where
67	15	alfo	all fo
69	29	one	our

Plate X. read the Characters of the Planets in this Order ♃ ☿ ♄ ♂ ♀

THE

A
LIST
OF THE
SUBSCRIBERS.

A.

*L*ORD Anſon.
   Hon. Mr. Archer.
Charles Ambler, *Eſq*;

B.

Duke of Beaufort.
Duke of Bedford.
Dutcheſs of Beaufort.
Lord Berkely, *of* Straton.
Miles Barne, *Eſq*;
Lancelot Barton, *Eſq*;
Hon. Antoine Bentinck.
Hon. John Bentinck.
Norbone Berkely, *Eſq*;
John Brown, *Eſq*;
—— Blaman, *Eſq*;
Thomas Brand, *Eſq*;
J. Bevis, *M. D.*
Rev. T. Bonney, *A. M.*

C.

Counteſs of Cunengeſby.
Lord Cornwallis.
Lady Cornwallis.
Edward Cave, *Eſq*;
John Chamock, *Eſq*;
Hon. and Rev. Dr. Cowper.
Mr. Richard Chad.
Mr. Henry Chapell.
Iſ. Colepepper.
Mr. George Conyers.

D.

Rev. John Dealtary, *A. M.*
Mr. Samuel Dent.

F.

Charles Fitzrea Scudamore, *Eſq*;
Kean Fitzgerald, *Eſq*;
Thomas Fonnerau, *Eſq*;
Robert Rakes Fulthorpe, *Eſq*;
Mr. Samuel Farrant.
Mr. Paul Fourdrinier.

G.

Marchioneſs Grey.
Lord Glenorchy.
Francis Godolphin, *Eſq*;
Roger Gale, *Eſq*;
James Gibbon, *Eſq*;
Ralph Goward, *Eſq*;
Ralph Gowland, *Eſq*;
Ralph Gowland, *Junior, Eſq*;
Dr. Gregory.
Dr. Griffith.
Rev. John Griffith, *A. M.*
Rev. Middlemore Griffith.

H.

Lord Hardwick, *Lord High Chancellor of* Great-Britain.
Hon. James Hamilton.
Mr. Thomas Heath.
Mr. Thomas Holt.
John Hughes, *Eſq*;

Earl

I.

Earl of Jersey.
Richard Jackson, *Esq;*
Rev. Mr. Jones.

K.

—————— Knowles, *Esq;*
Dr. Kendrick.
Mrs. Kennon, 4.

L.

Lady Vicountess Limerick.
Sir William Lee, *Bart.*
William Lester, *Esq;*
Rev. Dr. Long, *Master of* Pembroke-hall, Cambridge.
William Lloyd, *Esq;*
Mr. Andrew Lawrence.

M.

R. J. Mead, *M. D.*
Richard Meyrick, *M. D.*
Owen Meyrick, *Esq;*
Pierce Meyrick, *Esq;*

N.

Duke of Norfolk.
Lord North.
Lord Bishop of Norwich.
Richard Nicholls, *Esq;*
Mrs. Norsa.

P.

Duke of Portland.
Earl of Pembroke, *&c.* 2
Countess of Pembroke, *&c.*
Lady Palmerston.
Robert Money Penny, *Esq;*
Sir Francis Pool.
Sir John Pool.
John Probyn, *Esq;*
Rev. Mr. Pierce.

Mr. Dominick Pile.
Mr. Powel, *of* Cambridge,

R.

Dutchess of Richmond, *&c. &c.*
James Ralph, *Esq;*
Allan Ramsey, *Esq;*
William Read, *Esq;* 2.
Henry Reveley, *Esq;*
William Reveley, *Esq;*

S.

Sir George Savile.
—————— Serle, *Esq;*
Rev. Dr. Smith, *Master of* Trinity College, Cambridge,
Miss Stonehouse.
William Symonds, *Esq;*
Mr. James Scot.
Mr. James Stephens.

T.

Lord Viscount Townshend.
John Temple, *Esq;*
James Theobald, *Esq;*
Charles Townshend, *Esq;*
Mrs. Mary Trevor.
Mr. James Thornton.

V.

Lord Viscount Villiers.

W.

Lady Frances Williams.
Miss Williams.
Miss Charlotta Williams.
Rev. Thomas White, *A. M.*
—————— White, *Esq;*
Charles Louis Wiedmarkter, *Esq;*
Mr. Ward.   Y.
Hon. Philip York.
Dr. Arthur Young, *Preb. of* Cant.

LETTER the FIRST.

Opinions of the most eminent Authors whose Sentiments on the following Subject have been published in their Works.

SIR,

EFLECTING upon the agreeable Conversation of our last Meeting, which you may remember chiefly turned upon the Stars, and the Nature of the planetary Bodies ; a Subject, which is generally allowed to give true Pleasure to all those who take Delight in mathematical Enquiries ; and having not a little Regard to the repeated Request in your late Letters, I have at length undertaken to explain to you, as far as I am able, my Theory of the *Universe*, and the Ideas I have form'd of the known Creation.

The Hypothesis upon which this new Astronomy is founded, and now reduced into a regular System, was the result of my Astronomical Studies * full fifteen years ago, hence I hope you will allow, I have more than observed *Horace*'s celebrated Aphorism,

Nonumque prematur in annum.

* The first Scheme of this Hypothesis was plann'd in the Year 1734, representing in a Section of the Creation, eighteen Feet long and one broad, several thousand Worlds and Systems, and a great Number of emblematical Figures, now in the Author's Possession, together with a Scheme of the entire Creation, completed since, nine Feet long and six broad, more fully illustrating upon the same Construction the Innumerability of Systems and Worlds.

B   The

The Subject, I have often obferved, you have liftened to with a pleafed Attention, and I am the more incouraged to explain it at large to you, as I am perfwaded you don't want to be convinced of its valuable Ufes and Importance.

I remember you have often told me, that to apply ourfelves to the Study of Nature, was the fureft and readieft Way to come at any tolerable Knowledge of ourfelves, however difficult the Tafk might prove either in the Attempt, or the attaining it, and the lefs to be neglected, as it never fails to introduce a proper Knowledge of the DIVINE BEING, as a certain Confequence along with it, and fuch a Knowledge, as will naturally make every Man, who has but a tolerable Share of common Senfe, and is not a Slave to another's Reafon, without any other Evidence or Motive, in all Stations, and under all Circumftances, ACT JUSTLY, LIVE CHEARFULLY, and DIE full of Hope in the Expectation of a happy Sequel, in Futurity.

> *Eternity* is written in the Skies:
> Mankind's Eternity, nor *Faith* alone;
> *Virtue* grows there ————
>
>   *Dr.* YOUNG.

A learned Author on the Attributes, recommending thefe Studies as a reafonable and moral Service, fays, " Sure, it is moft becoming fuch imperfect Creatures as we are, to contemplate the Works of God with this " Defign, that we may difcern the Manifeftations of Wifdom in them; " and thereby excite in ourfelves thofe devout Affections, and that fu- " perlative Refpect, which is the very Effence of Praife."

> Who turns his Eye, *on Nature's Midnight Face,*
> *But muft enquire* ———— what Hand behind the Scene,
> What ARM ALMIGHTY, put thefe wheeling Globes
> In Motion, and wound up the vaft Machine?

The enchanting Idea *Milton* had of the Subjects of Aftronomy (whofe truly fublime Way of thinking and writing perhaps was never fo nearly equalled, or attempted before this Reverend Author's *Night-Thoughts,* appear'd is finely fhewn in the Eighth Book of his *Paradife Loft,* where he makes his *Adam,* fo earneftly attentive to the Angel *Gabriel,* as to ceafe relating the Myfteries of Creation.

The

The Angel ended, and his *Adam*'s Ear
So charming left his Voice, that he awhile
Thought him ftill fpeaking; ftill ftood fix'd to hear.

Milton's own Ideas of the Univerfe too, which no doubt he had ga-thered from aftronomical Authors, and had reconciled himfelf to, we are fully made acquainted with in the fame Book, where the Arch-angel fays, in anfwer to *Adam*'s Enquiries.

—— Other Suns perhaps
With their attendant Moons thou wilt defcry
Communicating Male and Female Light,
Which two great Sexes animate the World,
Stor'd in each Orb, perhaps with fome that live:
For fuch vaft Room in Nature, unpoffeft
By living Soul, defert and defolate,
Only to fhine, yet fcarce to contribute
Each Orb a Glimpfe of Light, convey'd fo far
Down to this habitable, which returns
Light back to them, is obvious to Difpute.

But before I prefume to plan my own Difcoveries and Conjectures into a Theory, both in Juftice to thofe who have in fome meafure been in the fame Way of Thinking, and alfo as a Defence of myfelf for producing fo new an Hypothefis to the World, which otherwife (though any Apology made to you I know will be unneceffary) may appear to too many but an idle *Chimera* of my own. I judge it will be highly proper, by way of ftrengthening my own Arguments, and adding more Weight to what I fhall myfelf advance in the following Letters, to give you in this the Opinions of the moft able Writers, whofe Works I have read upon the Subject. I mean fo far as relates to the now general received Notion, that the Stars are all Suns, and furrounded with planetary Bodies, with which I fhall fet out; and fhew you, it is not a Thing merely taken for granted, but has ever been the concurrent Notion of the Learned of all Nations, as fhall be further fhewn, in its proper Place, and as nearly as Poffibility will admit of, demonftrated to be Truth.

The following is an Extract from Mr. *Toland*, in his Account of the Works of

JORDANUS BRUNO.

" The Divine Efficacy (fays this Author in his infinite Creation) cannot
" ftand idle, without the Want of Will or Power; but any Imbecillity in

" such a Being argues Imperfection, and since any finite Produce com-
" pared with Infinity is as nothing, or rather as the Beginning of Good,
" it muftbe no lefs idle, and invidious in producing a finite Effect, than in
" producing none at all.

" Hence, as all Finites, fingly confidered, are but as Commencements
" of fomething more to be expected.

" Omnipotence, in making the Creation finite, will appear to be no
" lefs blameable for not being willing, than for not being able, to make it
" otherwife; i. e. infinite, as being an infinite Agent upon a finite Subject,
" which is repugnant to Reafon."

It follows then that, Creation muft be not only extenfively, but inten-
fively indefinite, and beyond the Reach of the human Underftanding to
comprehend; and that the one is as neceffary as the other, i. e. an infinite Expanfe is as reconcileable to our Reafon, as infinite Parts are to our
Senfes.

All the Attributes of the Divine Being are, as any one of them, incom-
fible to his Creatures; why fhould our Imagination then be fuppofed to
extend beyond the divine Activity?

" Thus, adds the above Author, the Excellency of God is adequately
" magnified, and the Grandeur of his Empire made manifeft; he is not
" glorified in one, but in numberlefs Suns; not in one Earth, or in one
" World, but in ten thoufand thoufand of infinite Globes."

An infinite Reprefentation of an infinite Original, and a Spectacle befit-
ting the Excellency and Eminence of him, that can neither be fully conceived, imagined, or comprehended.

> What read we here? th'Exiftence of a God?
> Yes, and of other Beings, Man above,
> Natives of Æther! Sons of higher Climes!
>
> Dr. YOUNG.

" If the Exiftence of this one World be good or convenient, it is not
" lefs good or convenient that there be infinite others like it.

" The infinite efficient Caufe would be abfolutely defective, without an
" infinite Effect; and befides, by conceiving the Infinity of the Univerfe
" and innumerable Beings, the Underftanding refts fatisfied, and is recon-
" ciled with the Idea of an Eternity; whereas, by afferting the contrary,
" it is unavoidably plunged into innumerable Difficulties, and unfolvable
" Inconveniencies, Paradoxes, and Abfurdities.

Again, fays the fame Writer, " Did we but confider and comprehend
" all this, oh! to what much further Confiderations and Comprehenfions
" fhoud

32

" fhould we be carried ! as we might be fure to obtain that Happinefs
" by virtue of this Science, which *in other Sciences is fought after in vain.*

> This Profpect vaft, what is it ? weigh'd aright,
> 'Tis Nature's Syftem of Divinity,
> And every Student of the Night infpires.
>
> Dr. Young,

> 'Tis elder Scripture, writ by God's own Hand;
> Scripture authentic ! uncorrupt by Man.

" This then is that Philofophy, which opens the Senfes, which fatisfies
" the Mind, which enlarges the Underftanding, and which leads Man-
" kind to the only true Beatitude, whereof they are capable according to
" their natural State and Conftitution ; for it frees us from the follicitous
" Purfuit of Pleafure, and from the anxious Apprehenfions of Pain, mak-
" ing us to enjoy the good Things of the prefent Hour, and not to fear
" more, than we hope from the future ; fince that fame Providence, or
" Fate, or Fortune, which caufes the Viciffitudes of our particular Being,
" will not let us know more of the one, than we are ignorant of the
" other."

And farther, " From thefe Contemplations, if we do but rightly confider,
" it will follow, that we ought never to be difpirited by any ftrange Ac-
" cidents, through Excefs of Fear or Pain, nor ever be elated by any prof-
" perous Event, through Excefs of Hope or Pleafure ; whence we have
" the Path to true Morality, and following it, we fhall of courfe become
" the magnanimous Defpifers of what Men of weak Minds fondly
" Efteem, and be wife Judges of the Hiftory of Nature, which would be
" written in our Minds, and confequently be chearful and ftrict Execu-
" tioners of the divine Laws, which would thus be ingraved in the Cen-
" ter of our Hearts. Seeking, as it were, in ourfelves, an Approbation of
" our own Action, which alone is capable of true Content and Happi-
" nefs."

CHRISTOPHER HUYGENS,

To whom the World is much indebted for many curious Inventions, and
Difcoveries, fays in his *Planetary Worlds*, " I muft be of the fame
" Opinion with all the great Philofophers of our Age, that the
" Sun is of the fame Nature with the fix'd Stars ; and this will give us a
" greater

---

† The Pendulum Clock; the firft Difcovery of *Jupiter*'s Satellites, and *Saturn*'s Ring.

" greater Idea of the World than all other Opinions can. For then
" why may not every one of thefe Stars, or Suns, have as great a Retinue,
" as our Sun, of Planets, with their Moons to wait upon them? Nay,
" there is a manifeft Reafon why they fhould; for, if we imagine our-
" felves placed at an equal Diftance from the Sun and fix'd Stars, we
" fhould then perceive no Difference at all betwixt them.

" Why then may we not make ufe of the fame Judgment that we
" would in that Cafe; and conclude, that our Star has no better Atten-
" dance than the others? So that what we allowed the Planets upon the
" Account of our enjoying it, we muft likewife grant to all thofe Planets
" that furround that prodigious Number of Suns. They muft have their
" Plants and Animals, nay, their rational Creatures too, and thofe as great
" Admirers and as diligent Obfervers of the Heavens as ourfelves; and
" muft confequently enjoy whatever is fubfervient to, and requifite for
" fuch Knowledge.

" What a wonderful and amazing Scheme have we here of the mag-
" nificent Vaftnefs of the Univerfe! So many Suns, fo many Earths, and
" every one of them ftock'd with fo many Herbs, Trees, and Animals,
" and adorned with fo many Seas and Mountains! And how muft our
" Wonder and Admiration be increafed, when we confider the prodi-
" gious Diftance and Multitude of the Stars?"

The Opinion of Sir ISAAC NEWTON.

This great Author, in his grand *Scholia* to the *Principia*, fays: — " The
" moft beautiful Syftem of the Sun, Planets, and Comets, could only pro-
" ceed from the Counfel and Dominion of an intelligent and powerful
" Being: And if the fix'd Stars are the Centers of other like Syftems, thefe,
" being form'd by the like wife Counfel, muft be all fubject to the Do-
" minion of One; efpecially, fince the Light of the fix'd Stars is of the
" fame Nature with the Light of the Sun, and from every Syftem Light
" paffes into all the other Syftems. And leaft the Syftems of the fix'd
" Stars fhould by their Gravity fall mutually on each other, he (the Di-
" vine Being) hath placed thofe Syftems at immenfe Diftances from one
" another."

The

The Opinion of Dr. DERHAM, *in his* Aſtro-Theology.

" The new Syſtem, ſays he, ſuppoſeth there are many other Syſ-
" tems of Suns and Planets, beſides that, in which we have our
" Reſidence; namely, that every fix'd Star is a Sun, and incompaſſed
" with a Syſtem of Planets, both primary and ſecondary, as well as ours.

" Theſe ſeveral Syſtems of the fixed Stars, as they are at a great and
" ſufficient Diſtance from the Sun and us; ſo they are imagined to be at
" as due, and regular Diſtances from one another : By which means it is
" that thoſe Multitudes of fixed Stars appear to us of different Magnitudes,
" the neareſt to us large; thoſe farther and farther, leſs and leſs; and
" that ſome, if not all of thoſe vaſt Globes of the Univerſe, have a Mo-
" tion, is manifeſt to our Sight, and may eaſily be concluded of all, from
" the conſtant Similitude and Conſent that the Works of Nature have
" with one another."

To this we may add, that this Syſtem of the Univerſe, as it is phyſi--
cally demonſtrable, is far the moſt rational and probable of any. *Becauſe,*

" It is far the moſt magnificent of any, and worthy of an infinite
" CREATOR, whoſe *Power* and *Wiſdom*, as they are without Bounds and
" Meaſure, ſo may they, in all Probability, exert themſelves in the Creation
" of many Syſtems as well as one.   And as Myriads of Syſtems are more
" for the *Glory* of GOD, and more demonſtrate his *Attributes* than one;
" ſo it is no leſs probable than poſſible, there may be many beſides this
" which we have the Privilege of living in."   And as the ſtrongeſt Confirmation of this, " we ſee it is really ſo, as far as it is poſſible it can be
" diſcerned by us, at ſuch immenſe Diſtances as thoſe Syſtems of the fixed
" Stars are from us; and we cannot reaſonably expect more."

" Since the Sun and fix'd Stars, ſays Dr. *Gregory*, are the only great
" Bodies of the Univerſe that have any native Light, they are juſtly
" eſteemed by Philoſophers to be of the ſame Kind, and deſigned for the
" ſame Uſes; and it is the Effect of a Man's Temper that ſets a greater
" Value upon his own Things than he ought, that makes him judge
" the Sun to be the biggeſt of them all."

That, as an elegant * Writer obſerves, which we call the Morning, or
the Evening Star, is, in reality, a *Planetary World*; which, with the four
others, that ſo wonderfully, as *Milton* expreſſes it, " vary their myſtick
" Dance, are in themſelves dark Bodies, and ſhine only by Reflection;
" have Fields and Seas, and Skies of their own; are furniſhed with all
" Accommodations for animal Subſiſtence, and are ſuppoſed to be the
                                                                    Abodes

* Contemplations on the ſtarry Heavens.

" Abodes of intellectual Life. Again, The Sun, with all its attendent Planets
" is but a very little Part of the grand Machine of the Universe.   Every
" Star— is really a vast Globe, like the Sun, in Size and in Glory, no less
" spacious, no less luminous, than the radiant Source of our Day ; so that
" every Star is the Center of a magnificent System, has a Retinue of
" Worlds irradiated by its Beams, and revolves round its active Influence ;
" all which are lost to our Sight in immeasurable Tracts of Æther.

" Could we, says the same Author, wing our Way to the highest ap-
" parent Star — we should there see other Skies expanded, other Suns,
" that distribute their inexhaustible Beams of Day ; other Stars, that gild
" the alternate Night ; and other perhaps nobler Systems established ;
" established in unknown Profusion, through the boundless Dimensions
" of Space.   Nor does the Dominion of the great Sovereign end *there*,
" even at the End of this vast Tour, we should find ourselves advanced
" no farther than the Frontiers of Creation ; arrived only at the Suburbs
" of the great *Jehovah*'s Kingdom."

> O for a Telescope his Throne to reach !
> Tell me ye Learn'd on Earth ! or Blest above !
> Ye searching, ye *Newtonian* Angels ! tell,
> Where your great Masters Orb ? His Planets where ?
> Those conscious Satellites, those Morning Stars,
> First-born of *Deity* from central Love.
>
> Dr. YOUNG.

Many other Authorities might be produced from Writers of great Repute, were it necessary to trouble you with them † ; but I believe those
above will be abundantly sufficient for the present Purpose, if even an
Apology were wanting for my own Conjectures.   I shall therefore conclude this Letter with the following Passage out of *Pope's universal Prayer*,
and in my next shall proceed in the Work I have undertaken.

> Yet not to Earth's contracted Span,
>   Thy Goodness let me bound ;
> Or think thee Lord alone of Man,
>   When thousand Worlds are round.
>
> *I am*, &c.

LETTER

† Particularly from *Fontenelle*, &c.

LETTER the SECOND.

Concerning the Nature of Mathematical Certainty, and the various Degrees of Moral Probability proper for Conjecture.

S I R,

YOU know how much I am an Enemy to the taking of any thing for granted, merely becaufe a Perfon of reputed Judgment, has been heard to fay, *it abfolutely is fo*; an *Ipfe dixit*, and implicit Faith in fome Cafes, may be both neceffary and ufeful; but here, in Aftronomy, I mean, every Man's Reafon, by the Help of a very little Mathematicks, is able to bring wonderful Truths to Light without them; and Truths not only of the higheft Importance to every Individual, but of a great and common Confequence to all Mankind : And as fuch, in all Ages of the World, have been judged worthy to be enquired into, by the beft and wifeft of Philofophers.

You are likewife very fenfible how far the human Underftanding is even at the beft, from being infallible, and don't want to be told, how difficult it is in a Subject of this Nature to arrive at any tolerable Degree of Certainty, which before the Days of the fagacious *Euclid*, and the penetrating *Archimedes*, was a Thing not to be expected. And many things which were then but barely Objects of Conjecture and Probability, have fince been demonftrated to be infallibly true. Time and Obfervation will undoubtedly, at laft, difcover every thing to us neceffary to our Natures, and proper for us to know. As a Proof of which, we fee human Wifdom daily increafes; and while a Capacity continues to make ourfelves ftill more acquainted with the manifeft Wifdom and Power of GOD in the Works of his Creation, who is to tell us where to ftop our Enquiries? Or who is fo impious to fet Bounds to a Science, which fo evidently fpreads through all Infinity, the Attributes of God, and an eternal Bafis for future Hope?

This Branch, or rather Body of Aftronomy, I believe you will find to be quite new; and though evident Truths, are the principal Thing to be regarded in it, yet as being in its infant State, where lineal Demon-

C   ftration ſtration fails, as in ſome Caſes it cannot be otherwiſe, I hope you will give me Leave to make uſe of a weaker Way of Reaſoning, to convince you of the Point in Diſpute, I mean of that by the Analogy of known and natural Things.

I ſhall be extremely unwilling to affirm any thing for a *Fact*, or Truth, without bearing, if not the real Evidence, at leaſt a plauſible Reaſon, next to a Conviction, or moral Certainty, along with it; and therefore I will here endeavour to explain to you what I mean by moral Certainty and alſo by mathematical Proof.

Mathematical Proof, or Certainty, proper for Conjectures, may, to almoſt every Capacity, be illuſtrated as follows:

Suppoſe you had accidentally found a very ſmall Part of a viſibly broken Medallion, with nothing more expreſs upon it, than what is repreſented at *Fig.* 1. *Plate* I. a Perſon totally unacquainted with the mathematical Sciences, we may naturally conclude, would not be able to make any thing of it, or in the leaſt comprehend what it originally was, or meant; but if an Aſtronomer ſhould chance to ſee it, who of courſe we are to ſuppoſe knew the Order and Proportion of the planetary Orbits, he would immediately conclude, and with great Probability, on the Side of his Conjectures, that it might be Part of a Medal repreſenting the Solar Syſtem. In ſuch a Caſe may we not very naturally ſuppoſe he would reaſon thus?

The Arches A and B ſeem to be Portions of the reſpective Orbits of *Saturn* and *Jupiter*, and what may lead us to believe, that they are really ſo, and Part of the Solar Syſtem, is the oblique Curve C, which looks not unlike the Trajectory of a Comet.

This ſurely would be far from an irrational Conjecture, and conſequently in ſome Degree probable: But this is not ſufficient you'll ſay; To prove it we muſt have farther recourſe to the Mathematicks, and a Mathematician would immediately thus demonſtrate it to be true.

Firſt, by compleating the Circles geometrically from the fourth Book of *Euclid*, by the Aſſiſtance of any three Points E. F. G. the original Figure will be reſtored, as at *Fig.* 2. And ſecondly, by aſſuming any two Points, as F, E in the Curve C, if admitted a Parabola, by a well-known Problem in Conic Sections the Heliocentric Portion X. Y. Z. will eaſily be projected and ſhewn, as in *Fig.* 3. Laſtly, join this in Poſition to the former, and it will juſtly ſupply the Orbit, or Path of ſome one of the Comets; and if required, even what Comet may be diſcovered by comparing the Perihelion Diſtance Y. S. with their general Elements or Theories, in Dr. *Hally*'s *Synopſis* of the Motion of theſe Bodies. And if a farther Confirmation of the Truth of theſe Conjectures were wanting,

the

PLATE.I.

Fig. 1.

Fig. 2.

Fig. 3.

PLATE II.

the small concentric Circles at D would now be allowed beyond a Contradiction, to represent the secondary Orbits of *Saturn*; and thus the first Presumption being carried thro' several corroborating Degrees of Probability, almost past a Dispute, would become a mathematical Certainty; and the above imperfect Piece of Medallion, would evidently appear beyond a Contradiction to be Part of a Representation of the said solar System, and such as is shewn in *Plate* II. Q. E. D. Thus in many Cases, it often happens, that from a very small Part of *orbicular Things*, we are able to determine the Form and Direction of the Whole: And hence you may conceive it no very difficult Task to a Mathematician, to describe the Orbits of all the Planets in the solar System, though he had never observed them but in one and the same Sign of the *Zodiack*; thus far I have thought it would not be amiss to explain to you the Nature of those Steps, by which we arrive at moral Certainty, and where the Subject will admit of it, Mathematical Conviction, which will not a little contribute to strengthen many of the Arguments hereafter made use of, and in some Degree serve to supply the Place of Proof, where infallible Demonstration cannot from the Nature of the Thing be discovered.

But besides the indisputable Principles of Geometry, the universal Law of Analogy and Similitude of things, have a Privilege to assist us, in Conjectures relating to the heavenly Bodies, and though not of equal Force with the former, is often as conclusive as the Subject requires. This sort of probable Evidence (as Dr. *Butler* observes,) is essentially distinguished from " Demonstrative by this, that it admits of Degrees; and " of all Variety of them, from the highest moral Certainty to the very " lowest Presumption; and that which chiefly constitutes Probability, is " expressed in the Word *Likely*, or Natural Likeness, as to State or Being." This general Way of arguing, I think, is allowed to be evidently natural, just and conclusive, and unquestionably to have its Weight in various Degrees, towards determining our Judgment: For Instance, should any ignorant Person, endowed with rational Principles, cut open a *Pomegranate* of the natural Growth of *England*, and finding it full of small Globules, or Kernels, upon being presented with an every way similar Fruit, said to be the Produce of *Italy*, doubt of its being of the same Nature, and composed of like globular Seeds within; here indeed would be no mathematical Evidence to assist the Judgment, the Object of Proof being invisible, but sure from the external Similitude, the strongest Probability of their being also internally the same.   Again,

Is it natural to suppose, that the first Person who found a *Lark*'s Nest, and in it several of the Female's Eggs, should have any Apprehensions of finding none in the *Nightingale*'s, only because he had never seen one before,

fore,

fore, I believe the moſt illiterate Perſon of the earlieſt Ages, who had Curioſity enough for ſuch a Search, would be greatly diſappointed in ſuch a Caſe; and far from concluding that the *Nightingale* had none. Farther, ſhould any one who had ſeen ſeveral Sorts of Fiſh taken out of the River *Thames*, or out of the *Nyle*; have any ſort of Suſpicion that he ſhould find no ſuch Creatures in the *Seine* or the *Ganges*, though it ſhould be allowed that he had never ſeen any ſuch Creatures that were known to come from thence. Ocular Demonſtration, in ſuch a Caſe, would ſure be unneceſſary, and an Evidence of the firſt, I believe would be abundantly ſufficient to convince us of what we ought to look for at leaſt in the laſt: But then the Fiſhes of different Seas, and of Rivers are not of the ſame Speeies you'll ſay; but as it were infinitely diverſified through all the aqueous World, this is, and muſt be granted, and alike Variety of *Species* muſt alſo be granted, in the former Caſe of the Birds: But no Objection can poſſibly ariſe from any ſuch Diverſity, ſince we don't pretend to ſay, nor is it at all neceſſary, that the Beings in the ſidereal Planets ſhould be every where the ſame with theſe of our ſolar Syſtem, a Variety muſt every where be admitted, and will always be admired, where the Work is Nature's, and the Deſign God's.

All then that I here pretend to argue for, is a Univerſality of rational Creatures to people Infinity, or rather ſuch Parts of the Creation, as from the Analogy and Nature of Things, we judge to be habitable Seats for Beings, not unlike the mortal human.

Every Animal, and every Vegetable, that, as it were, naturally exiſts by the Virtues, Properties, or Laws of the mineral Kingdom, has ſomething of a ſecondary Nature, depending upon it as a Principle; and to ſay that the Stars, which are a certain viſible ſort of Cotemporaries in Space with the Sun, have no like planetary Bodies with ours moving round them, becauſe we cannot poſſibly ſee them, is no leſs abſurd and ridiculous, than to argue, that we can have no Reaſon to expect to find, in the proper Seaſon, Grapes upon every Vine — Figs upon every Tree — Roſes upon every Buſh — only becauſe ſome of them are at ſuch a Diſtance, that neither Roſe, Fig or Grape, can be diſcovered by the Eye.

This ſort of Reaſoning, though ſome perhaps may neglect it, I am perſwaded you will look upon as abundantly ſufficient for Things out of the Reach of Science to determine; and that the collective Body of Stars have not been diſcovered, to be together a proper Subject for ſuch Conjectures before, can ſurely only proceed from the Want of Time, neceſſary to compleat the Obſervations proper for a Foundation to build ſuch an Hypotheſis, or Theory upon. This is the great Article in which the Moderns have ſo much, and ever will have, an Advantage over the Antients. And hence it will appear, That

The

44

The Improvements and Difcoveries of latter Ages are not at all owing to the greater Capacity of the Moderns, but from the Advantages receiv-ed, or arifing from the Inventions and Progrefs made by the Ancients. We at firft in a manner walked by their Leading-ftrings, and though many of them now are broke, or ufelefs, none can deny, but that formerly they were of great Advantage in promoting and directing philofophical En-quiries.

In an Affembly of the moft eminent Men of all Ages, if we may fup-pofe fuch a Conference amongft the illuftrious Dead, on Purpofe to deliver their feveral Sentiments familiarly together, on the moft interefting Sub-jects of natural Knowledge, who would not lament the Difadvantages, poor old *Thales*, an *Hipparchus*, or a *Ptolomy*, would lie under, who had no-thing but the Eye of Reafon to direct them, in Oppofition to the Judg-ment of a *Brahe*, or a *Galilæus*, who reaped fo much Benefit from their compound Opticks? But on the other hand, perhaps if the folar Syftem, was the Topic of Difcourfe, a * *Pythagorean* might very pertinently fay to a *Newtonian*, " You have not gone much farther in the Light with our " Direction, than we did in the Dark alone; for you are ftill roving " round the fame Circles." Much might be faid upon this Head; but I believe it would be a difficult Matter to do Juftice to all Parties: So here I intend to leave them, only muft obferve, that Pofterity will always have the Advantage over their Predeceffors; and that After-ages, in all Pro-bability, will reap fo great a Benefit from the Invention and Improve-ment of Fluxions, that fcarce any thing, which is the immediate Ob-ject of fuch Enquiry, will long lie concealed from a true mathematical Genius.

For this, in which he has furpaffed all the Antients, and greatly advanced the philofophical Sciences, the World is indebted to Sir *Ifaac Newton*.

But as many of his Difcoveries, fuch as relate particularly to the Laws of the planetary Syftem, are but as fo many Confirmations of the Con-jectures and Imaginations of Aftronomers and Philofophers before him, it perhaps will not be amifs to acquaint you a little with the Aftronomy of the Antients concerning the Univerfe. And before I proceed to thofe of my own, fhew you in the firft Place how far their Speculations in the vi-fible Creation have been carried; and with thefe I fhall conclude this pre-paratory Epiftle.

The Univerfe, or mundane Space, by which the Antients comprehend all Creation, has, from time to time, according to the Progrefs of Science, come under a fort of Neceffity of being varioufly modell'd agreeable to the

Opinion

---

* The true Syftem of the Planets have been difcovered above two thoufand Years.

Opinion of the feveral Authors, who have judged themfelves wife enough to write upon it with a mathematical Foundation; And the cofmical Syftem, by which is meant the Co-ordination of its conftituent Parts has undergone almoft as many Changes as its Elements are even capable of ; every Age of the World, as Knowledge has increafed, either from improved Imagination, or repeated Obfervations, producing fomething new concerning it.

MILTON, no doubt, had all this Diverfity of Opinions in View, as appears from his fuppofed Pre-knowledge of *Raphael*, in the following' Paffage, *Book.* VIII.

> Hereafter, when they come to model Heaven,
> And calculate the Stars, how they will weild
> The mighty Frame ! how build, unbuild, contrive
> To fave Appearances, how gird the Sphere
> With centric and eccentric fcribbl'd o'er;
> Cycle, and Epicycle, Orb in Orb.

But the following Synopfis, I believe, will abundantly convince you that from certain Obfervations only, we ought to form all our Notions of it, if we either hope to arrive at Truth, or expect our Ideas fhould be fupported by Reafon.

ARISTOTLE was of Opinion, that the Univerfe, or Heaven, was all one World, and St. CHRYSOSTOM, TERTULLIAN, St. BONAVENTURE, TYCHO BRAHE, LONGOMONTANUS, KEPLER, BULIALDUS and TELLEZ, were of an united Opinion, that this one Heaven, or Univerfe, was all fidereal and fluid. But AEGIDIUS, HURTADUS, CISALPINUS, and AVERSA, believing the fame Heaven with them to be all one World, and that fidereal, yet on the contrary held it to be folid.

CLEMENS, ACACIUS, THEODORET, ANASTASIUS, SYNAITA, PROCOPIUS, SUIDUS, S. BRUNO, and CLAUDIANUS MAMERTUS, fuppofed the univerfal mundane Space as divided into two Heavens, namely,

The Empyræum created the firft Day,

And the Firmament created the fecond Day.

Two Heavens were alfo held by JUSTIN MARTYR, the one fidereal, and the other aerial. The firft fuppofed by St. GREGORY NYSSENE, to be that of the fixed Stars, and the laft, that of the Planets. But *Maftrius* and *Bellutus*, though agreeing in the Number of Heavens, call one the *Primum Mobile*, and the other, the Starry Heaven.

Farther,

Farther, St. BASIL, St. AMBROSE, DAMASCENE, CASSIODORUS, GE-NEBRARDUS, SUAREZ, TANNERUS, HURTADUS, OVIEDUS, TELLEZ, and BORRUS, diftinguifhed the Univerfe as divided into three Portions, or Heavens.

The firft called the Empyræum,	*Or, as Cajetan.*	Watery,	*Tho. Aquinas.*	
The fecond fuppofed Sidereal,		Sidereal,		Watery,
And the laft of all, Aerial.		Aerial,		Sidereal.

Again, St. *Athanafius* adds to thofe of the fix'd Stars, the Planets, and the Air, that of the *Empyræum*, and makes in all four Heavens.

But as the Number of the Heavens thus increafes, and will become fubdivided in the fubfequent Account of them, to give you a better Idea of the Order of thefe celeftial Portions of the mundane Space, it will not be amifs to form what remains of them into regular Sections of their proper Spheres and Syftems.

See *Plate* III. in which Figure, the firft reprefents a Section of the cofmical Theory of *Oviedus* and *Ricciolus*: Both confifting of five Heavens, *viz.*

By Oviedus, sidereal and folid.			*By Ricciolus,* fider. and fluid.		
The fixed Stars,	A			*Empyræum*, - - G	
Saturn, - - - - - -	B			The Water, - - F	
Jupiter, - - - - -	C			The fixed Stars, A	
Sol, with ♂, ☿ and ♀ included	D			The Planets, - H	
The Moon. - - - -	E			The Air. - - I	

Fig. II. reprefents that of venerable *Bede* and *Rabanus*, *viz.* of Seven Heavens.

And according to *Bede* compofed of		But by *Rabanus*,
The Air, - - — - -	P	The Atmofphere,
The Æther - - - - -	O	The upper Air,
Olympus, - - - - - -	N	The inferior Fire,
The Element of Fire, - - -	M	The fuperior Fire,
The Firmament, - - - -	A	Sphere of the fixed Stars,
The Angelical Region, - -	L	The Chryftalline Heaven,
Realm of the Trinity. - -	K	The *Empyræum*.

Fig.

47

Fig. III. Reprefents the Hypothefes of *Eudoxus, Plato, Galippus, Cicero, Riccius, Philo, Remigius, Aben-Ezra, Carthufianus, Lyranus, Toftatus, Brugenfis, Orontius, Cremoninus, Philalethæus, Amicus,* and *Ruvius*; alfo the *Babylonians* and *Egyptians.*

Confifting of Eight Heavens,

All Sidereal, *viz.* The Sphere of the fix'd Stars, and thofe of the Seven Planets.

Fig. IV. is that of *Macrobius, Haly Alpetragius, Rabbi-Jofue, Rabbi Moyfes, Scotus, Abraham Zagutus, Sacrobofcus, Claromontius, Avigra,* and *Arraiga.*

All of Nine Heavens,

Comprehend a *Primum Mobile* Q, or, according to *Arriaga,* a folid *Empyræum.* The Sphere, of fixed Stars A, and the feven Regions of the folar Planets.

Fig. V. is that of the great *Alphonfus, Fernelius, Regiomontanus, Amicus, Maurolycus* and *Langius*; alfo of *Azahel, Thebit,* and *Ifaac Ifraelita*; and likewife of *Gulielmus Parifienfis,* and *Johannes Antonius Delphinus.*

Confifting of Ten Heavens, made up of

A *Primum Mobile*	— — —	S	*Empyræum.*
A Sphere of *Tripidation* in Longitude	—	R	*Primum Mobile.*
The Sphere of the fixed Stars	— —	A	

And thofe of the feven folar Planets within.

Note, Some Authors place the Sphere of *Tripidation* in Longitude below that of the *Aplain,* or Eighth Sphere.

Laftly, Fig. VI. is the Heaven of *Petrus Alliacenfis,* the College of *Conimbra, Martinenfis,* (and fometime) of *Clavius*; and alfo *Johannes Warnerus, Leopoldus de Auftriâ, Johannes Antonius Maginus*; and laftly, of *Clavius.*

In all Eleven Heavens containing,

T　A *Primum Mobile,* or, as others fay, an *Empyræum.*

V　A Sphere of Libration in Latitude.

W　A Sphere of Libration in Longitude.

A　The Sphere of the fixed Stars, and thofe of the Planets.

Thus you fee how many various Opinions have from time to time been imbraced concerning the Fabric and Formation of the vifible Univerfe; all of which are now and have long been exploded; and although at firft advanced by Men of the greateft Learning, and of the deepeft Penetration in natural Knowledge, it does not appear from any one of their Opinions, that they had any the leaft Notion of infinite Space, but as it

were

PLATE III.

Fig. I.

The Seven PLANETARY ORBITS.

were confined the Divine BEING to their limited Notions, as one may say in an Egg-fhell. If therefore what I fhall hereafter advance, extend fo far without the known Creation, that you can poffibly conceive no Bounds to the Works of infinite Wifdom and Power, I hope you will be in no Danger of looking upon it as more ridiculous, or abfurd, than what fo many of the wifeft Men of every Age have thought proper to attempt, and have judged worthy of their Attention fo long before me. If any thing lefs fo, I fhall think myfelf happy enough in having broke, or rather paffed the narrow Limits to which the Creation has for fo many Years been confined, in hopes of tempting Men of greater Talents to look upwards, and purfue fo noble a Subject as far as the human Underftanding is capable of comprehending it.

To the Opinions above might be added many more, particularly that of *Johannes Baptifta Turrianus*, and *Fracaftorius*, who increafed the Number of Heavens to fourteen, *viz.* feven on each Side the *Aplané*.

But of this I have faid enough; in my next I fhall proceed to Matter better grounded,

And am, &c.

LETTER

LETTER the THIRD.

Concerning the Nature, Magnitude, and Motion of the Planetary Bodies round the Sun, &c.

SIR,

THE younger *Pliny*, if I remember right, somewhere says, that there is, or ought to be, a wide Difference betwixt writing to a Friend, and writing to the Publick : I have indeed pleased myself with the one, but am far from thinking myself qualified for the other ; I must therefore rather intreat you, though perhaps you cannot possibly overlook all my Faults as an Author, to excuse them at least in the Friend, and by such kind of unlimited Indulgence, you will give me a much greater Chance to do the Subject some Justice, though I own I despair in this first Attempt, to reconcile every thing I advance to your more cool and impartial Reasoning. But to the Business :

As I have no Ambition to have the Substance of my Theory more admired by you than understood, which is too often the Case in Works of this Nature, I must beg leave to repeat to you Part of a former Discourse, which will refresh in your Ideas the principal Laws of the System of our Sun, and make you properly acquainted with such Things as are necessary to be known in the now-established Astronomy of * *Copernicus*, &c. before I proceed to any new Matter.

The

---

* NICOLAUS COPERNICUS, stiled by *Bulialdus*, *Vir absolutæ subtilitatis*, was a Native of *Thorn* in *Polish Prussia*, and Canon of the Church of *Frawenburgh* ; he was Scholar to *Dominicus Maria* of *Ferrara*, to whom he was Assistant in his astronomical Observations at *Bologne*, and Professor of the Mathematicks at *Rome*, in his noble Work, *De Revolutionibus Orbium Cælestium* ; he fortunately revived, happily united, and formed into an Hypothesis of his own, the several Opinions of *Philolaus*, *Heraclides Ponticus*, and *Ecphantus Pythagoreus*, viz. after the Opinion of *Philolaus* he made the Earth to move about the Sun, as the Center of its annual Motion ; and according to *Heraclides* and *Ecphantus*, he likewise gave it a diurnal Rotation round its own Axis : Which System has withstood all Opposition ; and as *Ricciolus*, (though a Dissenter from it) observes, *Per damna, per cædes, ab ipso sumit opes, animumque ferra.*

52

The Sun, you are not to learn, is the reputed Center of our *Planetary System*, and may remember, that the Earth on which we live, and thefe five following *Erratic Stars*, viz. SATURN, JUPITER, MARS, VENUS and MERCURY, have been demonftrated to move round him in the Order and Manner following.

Saturn is found to complete one Revolution round the Sun in twenty-nine Years, one hundred and feventy-four Days, fix Hours, and thirty-fix Minutes; at the Diftance of about feven hundred and feventy-feven Million of Miles. *Jupiter* performs a like Revolution in about eleven Years, three hundred and feventeen Days, twelve Hours, and twenty Minutes; diftant from the Sun about four hundred and twenty-four Millions of Miles. *Mars* compleats his Circuit in one Year, three hundred and twenty-one Days, twenty-three Hours, and twenty-feven Minutes; and his mean Diftance is about one hundred and twenty-three Millions of Miles.

Thefe three are called fuperior Planets, as being farther from the Sun than the Earth, and circumfcribing its Orbit.

The Earth circumambiates her Orbit in one folar Year, *viz.* in three hundred and fixty-five Days, five Hours, forty-eight Minutes, and fifty-feven Seconds; at the mean Diftance of eighty-one Million of Miles.

The Radius of *Venus*'s Orbit is about fifty-nine Millions of Miles; and that of *Mercury* nearly thirty-two Millions, *ditto*.

The Heliocentric Revolution of *Venus*, is made in two hundred and twenty-four Days, fixteen Hours, forty-nine Minutes, and twenty-feven Seconds; and that of *Mercury*, in eighty-feven Days, twenty-three Hours, fifteen Minutes, and fifty-four Seconds. Thefe two laft Planets are called inferior Ones, as being circumfcribed by the Earth.

The Diameter of the Sun being demonftrated to be nearly feven hundred and fixty-three thoufand Miles:

The proportional Magnitudes of all the above Planets will be found nearly as follows, *viz.*

The Diameter of the Globe,

Of *Mercury* - -	4,240	
Venus - -	7,900	
the Earth - -	7,970	} Miles.
Mars - -	4,440	
Jupiter - -	81,000	
and *Saturn* - -	61,000	

D 2

Thus

53

Thus much I have thought proper to premife, and for your immediate Infpection, have added the following Schemes, that nothing may be wanting to give a general Idea of the Order of the celeftial Bodies in our own Syftem, before I attempt to lead you through the neighbouring Regions of the Stars to the more remote Tracts of Infinity.

PLATE IV.

Is a true Delineation of the folar Syftem, with the Trajectories of three of the principal Comets, whofe Periods and Orbits have been accurately determined, and are reprefented in their true Proportion and Pofition to one another, and the Order of the Planets round the Sun, marked with their refpective Characters, *viz.* ♄, for *Saturn,* ♃, *Jupiter,* ♂, *Mars,* ⊕, the Earth, ♀, *Venus,* and ☿, *Mercury.* The Scale being nearly five hundred and eighteen Millions of Miles to an Inch.

PLATE V.

Is a true Projection of the Syftem of the known Comets; in which are reprefented nine of the chief Trajectories, from their *Aphelii* to their *Perihelii,* all in juft Proportion and Pofition to the Orbits of *Saturn* and *Jupiter,* which are alfo reprefented by the two concentric Circles, fuppofed to be drawn round the Sun as their Center.

The Ellipfis, or Trajectory, marked A, fhews the Pofition and Path of the Comet which appeared in the Year 1684, whofe Period is fuppofed to be about fifty Years, and has been obferved within the Region of the Planets once.

That mark'd B, is the Way of the Comet of 1682 ;
The Period conjectured to be about feventy-five Years and a half, and has been obferved thrice.

C, Way of the Comet of 1337 ;
The Period about 100 Years, obferved once.

D, That of the Comet of 1661 ;
The Period about 129 Years, obferved twice.

E, Tract of the Comet of 1618 ;
The Period about 160 Years, obferved once.

F, Way of the Comet of 1677 ;
The Period about 200 Years, obferved once.

G, Way of the Comet of 1744 ;
The Period about 300 Years, obferved once.

H, Way of the Comet of 1665 ;
The Period about 400 Years, obferved once.

I, Way of the Comet of 1680 ;
The Period about 575 Years, obferved thrice.

The

PLATE IV.

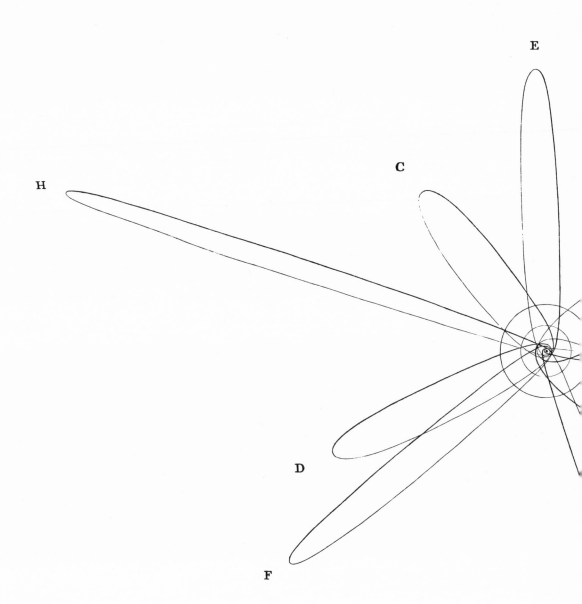

PLATE V.

B

A

G

I

The Scale of this Syftem is equal to one Third of the former.

Here I muft obferve to you, as a Thing I judge may prove of great Confequence with regard to the Syftem of Comets, which is as yet very imperfect: That I am ftrongly of Opinion, that the Comets in general, through all their refpective Orbits, defcribe one common Area, that is to fay, all their Orbits with regard to the Magnitude of their proper Planes, are mathematically equal to one another; which, if it once could be proved, and confirmed by Obfervation, the Theories of all the Comets that have been juftly obferved, might eafily be perfected, and their Periods at once determined, which now we can only guefs at, or may wait whole Ages for more Certainty of. What leads me to believe, that this may prove to be really the Cafe is this.

I find by Calculation, that the Orbits of the two laft Comets, whofe Elements have been moft corrected by Sir *Ifaac Newton* and Dr. *Hally*, are to one another, according to their Numbers,   nearly as * 13 to † 17, notwithftanding one of them is one of the moft erratick that ever came under our Obfervation ; and the other one of the moft neighbouring to the Sun.

But it is well known to all Mathematicians, that the firft of thefe Comets moved in fo eccentric a Trajectory, that the leaft Error in its almoft incredible Proximity to the Sun will produce a very fenfible Difference in the Area of the Orbit: And accordingly, if we moderate the Perihelion Diftance of this Comet, by making it but 1000 inftead of ‡ 612, which is but increafing it a $\frac{1}{35000}$th Part of the great Radius of the Orbit, (which is an Error every Aftronomer will readily grant is very eafily made) and we fhall find the Orbits of the faid two Comets to be exactly equal.

Further, I muft inform you, that the Comet of 1682, which the above compared with, feems to have been fo accurately obferved, that it does not appear to have altered its Perihelion Diftance half a 68th Part in one intire Revolution. Now, if we can with any Show of Reafon, and a Probability on our Side, bring the Areas of thefe two extream Comets, as I may call them, to an *Equality*, fure we may conclude, it is a Subject highly worthy to be more confidered and enquired into.

PLATE

* 1316539,968282 Comet of 1680.
† 1708155,4644 Comet of 1682.
‡ The Number in Dr. *Hally*'s Synopfis.

P L A T E VI.

Is a true Reprefentation of the fatellite Syftems, proportionable to one a-nother, and to the Orb of the Sun's Body, that a juft Idea of the Diftances of thofe fecondary Planets, may be eafier had from their refpective primary ones.

S reprefents the folar Body with its Atmofphere. *Fig.* 1. is the Syf-tem of *Saturn* from the fame Scale. *Fig.* 2. that of *Jupiter* from *ditto.* And *Fig.* 3. the Orbit of the Moon round the Earth, in the fame Proportion.

But as you can have but a very imperfect Idea of the Magnitude of thefe laft Circles, with regard to the Body of the Earth or Moon,

P L A T E VII.

Is a true Projection of their real Globes, at their proper Diftance from each other, with their common Center of Gravity, and the Point and Line of equal Sufpenfion betwixt them, *viz.*

A, reprefents the Globe of the Earth.

B, that of the Moon.

C, Point, and C D, Line of equal Sufpenfion betwixt them.

E, Common Center of Gravity, which defcribes the *Orbis Magnus.*

E, F, and B, G, is the Orbit of the Moon.

Farther, that nothing may be wanting to give a true Notion of the whole together,

P L A T E VIII.

Is a proportional Drawing of all the primary and fecondary Planets to-gether, diftinguifhed by their Characters, proper to attend a Globe of twelve Inches Diameter, fuch a one being fuppofed to reprefent the Sun.

P L A T E IX.

Is an exact Scheme of the principal known Comets, in juft Proportion, to the Globe of the Earth reprefented at A, with the Nuclus, and Part of the Tail of the Comet of 1680, B, as it was obferved in its Affent from the Sun, *viz. a a* the Comet's natural Atmofphere, *z z z,* the *Denfer Matter* winding itfelf into the Axis of the Train *x x,* the inflam'd Atmofphere and Tail dilated near the Sun. C, reprefents the Ball of the Comet of 1682, D, that of 1665, E, that of 1742, and F, the Head of the Comet of 1744.

And again, that you may have fome Notion of the apparent Magnitudes of all thefe Planets and Comets, *&c.* as they appear at the Earth,

P L A T E

PLATE . VI .

Figure . I .

S

Fig III

Fig . II .

PLATE . VII.

PLATE VIII.

PLATE X.

PLATE X.

Reprefents the Sun and Moon in the juft Proportion of their mean Diameters, with two of the Comets A and B, and the five erratick Planets, as they are obferved at the Earth, in a middle State of their Diftances from it.

For a more full and particular Defcription of all the Parts of the folar Syftem, and of the home Elements of Aftronomy in general, I refer you to my *Clavis Cœleftis*,&c. where every thing concerning the Planets, Comets, and Stars; and their real and apparent Motions, are at large reprefented, explained, and accounted for, for the Benefit of fuch as have not made the Mathematicks their regular Study.

Now, to convince you that the Planets are all in their own Nature no other than dark opaque Bodies, reflecting only the borrowed Light of the Sun, I muft recommend to your Obfervation, this natural and fimple Experiment, which almoft any Opportunity of feeing the *Moon* a little before the Full, will put into your Power to make; but beft and eafieft when the Sun is in any of the North Signs, *i. e.* in *Summer*.

At fuch a time, the Sun being near fetting, the Moon will appear in the eaftern Hemifphere; and if there be any bright Clouds northward, or fouthward near her, you will plainly perceive, that the *Light* of the one is of the fame Nature with that of the other; I mean the Light of the Moon, and that of the Cloud. To me there never appeared any Difference at all; and I am perfwaded, were you to make but two or three Obfervations of this kind, which is from Nature itfelf, a fort of ocular Demonftration, you cannot fail of being convinced, that the Moon's Light, fuch as it is, without Heat, can poffibly proceed from no other Caufe than that which illumines the Cloud: For if the Clouds, whofe Compofition we know to be but a thin light Fluid, formed of condenfed Vapours only, is capable of remitting fo great a Luftre, how much more may we not allow the Moon, which, Length of Time, and many other Circumftances, have long confirmed to be a durable and folid Body.

The Increafe of her Luftre, indeed, during the Abfence of the Sun from us, to a lefs penetrating Genius than your's, may poffibly afford fome trifling Ground of Objection to the above Conclufions, as being drawn from the Phænomena of Day-light only; by reafon in the Night, we have no Clouds in equal Circumftances to compare with her.

But this I need not tell you, is all owing to her being feen through a darker Medium, and not to any real Increafe of natural Light emitted from the Sun. As a Proof of which, were it neceffary, you need only, fhut out the Rays of the Atmofphere, by the Help of a fufficiently

long

73

long Tube; and the Moon, or any other celeftial Body, will appear through it, as bright in the Day-time as in the Night.

Thus all light Bodies of inferior Luftre, whether fhining by their own natural Radiences, or by a borrowed Reflection, partake of the fame Advantage, when removed from the more potent Influence of a fuperior one; and hence it is, that the *Aura Ætherea fhines out moft manifeft, when the Body of the Sun himfelf is hid, the Stars, and the *Via Lactea* moft lively and numerous in the Abfence of the Moon, and thofe Exhalations, or Meteors, vulgarly called Falling-ftars, become only vifible (like Glow-worms) in the Night.

Here it may not be improper to tell you, that the Clouds are to us in effect no other than as fo many Moons, whereby we have our artificial Day prolonged to us feveral Hours after the Sun is fet, and likewife produced as much fooner before he rifes; and were they to afcend by ftill ftronger Power of Exhalation to an Elevation, all round the Atmofphere, fo as to form a Sphere equal to four Times the Globe of the Earth, there would then be no fuch Thing as real nocturnal Darknefs to any Part of the World.

The lunar Light then we may very juftly conclude, proceeds originally from the Sun: And notwithftanding many more Arguments might be drawn from the Demonftration of her Phafes, Eclipfes, &c. to prove it, yet none of them need here be added, to what has been already faid, to convince you of the Truth of it. This being granted, let us now confider what Effect this, or a like Quantity of borrowed Light, would have, when removed to a much greater Diftance.

I may, I think, fuppofe, that you know fo much of Opticks as to underftand, that all vifible Objects apparently decreafe in Magnitude, as their Diftance from the Eye increafes. Confequently, that, if the Moon's Orbit was placed as far again from the Earth as it really is, her Globe, or rather *Difk*, would then feem to be but half as big as to us fhe now appears to be, and of courfe ftill farther, were fhe placed at ten times the Diftance fhe is known to revolve at, her apparent Diameter would be reduced to a tenth Part only of what it now appears to be in her prefent Orbit, that is, one hundred Times lefs in vifible Magnitude than her neighbouring Difk is found to be where it now is feen. And fuch, but fomething lefs, the two Planets *Venus* and *Jupiter*, which are frequently, in their Turns, our Morning and Evening Stars, appear to be through a common Telefcope.

Now

---

* An *Helios*, or golden Light, always attending the Sun, and fuppofed to fpread itfelf all round his Body in the Direction of his Equator, was very vifible during the total Darknefs of the Eclipfe of 1715, and may be always feen about the Autumnal Equinox.

Now thefe two Planets, together with the other three, which we find moving in regular Orbits round the Sun, are all found fubject to the fame * Changes of *Phænomena*, in their various Afpects with the Sun; and who can doubt but that they are all of the fame or like Nature? But you'll fay, perhaps, how are we fure that *Venus* and *Jupiter* have no native Light of their own, fince many of the ancient Philofophers, and in particular *Anaximander*, allowed even the Moon to have fome; and befides, in Philofophy, as well as in Logick, I think you hold there is no proving a Negative, at leaft at fuch a Diftance.

To make you conceive the Impoffibility of fuch a Light, and next to a Demonftration, convince you of the Unnaturalnefs of fuch a Suppofition, I muft put you in mind, that fome time ago, when I was laft in the Country with you, I think it was about the latter End of Autumn, near the Winfter Solftice, as we were walking one Evening, I bid you take notice of the Moon, which was then near fetting, and about two Days old. You may remember, her whole Globe appeared to us very confpicuoufly within a manifeft Circle. You immediately told me, that that kind of Phænomenon the Country People called a *Stork*, or the old Moon in the new one's Arms. This I then endeavoured to explain to you, and I think made you fenfible it was intirely an Effect of the Earth's, and an Appearance always to be expected at that Time of the Year. The Earth being then in the State of a Full-Moon to that Part of the lunar Orbit, and near her Perihelion, at which time, the Earth fends back a Reflection to the † Moon twenty-five times more potent than that of the Moon to us.

Now the Planet *Venus*, from undeniable Principles of Geometry, is allowed to be nearly fuch another Globe as the Earth is; and fince the Earth, as I have juft now related, is found to reflect much more Light to the Moon, by reafon of her fuperior Magnitude, than the Moon can poffibly reverberate to Earth again; and fince alfo 'tis plain, the Earth has no Light of its own, why then fhould we imagine *Venus* to be endowed with a Luftre, which we can prove to be no more than a fimilar Body, and governed by the fame Laws as the Earth is?

Anaximander's Miftake, in fuppofing the Moon in fome fmall Degree a radiant Body of itfelf, lay, in not confidering, that the faint Illumination here defcribed, and vifible all over her Globe, foon after almoft every Conjunction with the Sun; and probably in Eclipfes, alfo proceeded from the Earth; but the thing I think is too evident to expect any fort of Contradiction,

E

☞ * *Venus* and *Mercury* in every Heliocentrick Revolution, perform all the Changes of our Moon in a like Gradation and Defection of Light, both horned and gibos'd.

† Their Diameters being nearly as 1 to 5.

75

tradiction, therefore I hope you will admit it as a Truth, and confequently take it for granted, that the planetary Bodies in general, are meer terreſtrial, if not terraqueous Bodies, ſuch as this we live upon; which is the Thing I have chiefly in this Letter attempted to demonſtrate, or have rather explained; and now I hope, for the future, you will receive the Idea of a Plurality of Worlds more favourably, and look upon aſtronomical Conjectures in a leſs ridiculous Light than you uſed to do, eſpecially ſince you muſt allow, they give our unlimited Imaginations a like all endleſs Field of Contemplation, not only full of the wonderful Works of Nature, but alſo of a viſible Providence.

I think I cannot conclude this Letter to you more properly, than with the following fine Lines of Mr. *Addiſon*'s from the *Spectator*, Vol. VI. No. 465. which I hope you are not ſo polite as to look upon as an unfaſhionable Quotation.

The ſpacious Firmament on High,
With all the blue ethereal Sky,
And ſpangl'd Heav'ns, a ſhining Frame,
Their great Original proclaim:
Th' unwearied Sun, from Day to Day,
Does his Creator's Pow'r diſplay,
And publiſhes to ev'ry Land
The Work of an Almighty Hand.
Soon as the Ev'ning Shades prevail,
The Moon takes up the wond'rous Tale,
And nightly to the liſt'ning Earth,
Repeats the Story of her Birth:
Whilſt all the Stars that round her burn,
And all the Planets in their Turn,
Confirm the Tidings as they roll,
And ſpread the Truth from Pole to Pole.
What though, in ſolemn Silence, all
Move round the Dark terreſtrial Ball?
What tho' nor real Voice nor Sound
Amid their radiant Orbs be found?
In Reaſon's Ear, they all rejoice,
And utter forth a glorious Voice,
For ever ſinging, as they ſhine,
" *The Hand that made us is divine.*"

And am, &c.

LETTER

LETTER the FOURTH.

Of the Nature of the heavenly Bodies continued, with the Opinions of the Antients concerning the Sun and Stars.

S I R,

YOU tell me you begin to be a tolerable good *Copernican*, and would now be glad to have my Opinion further upon the Nature of the Sun and Stars, with regard to the Suggeſtion of their being like Bodies of Fire. This you ſay will go a great Way towards confirming you in the Notion you have begun to embrace of a Plurality of Syſtems, and a much greater Multiplicity of Worlds than our little ſolar Syſtem can admit of. Beſides, ſhewing in a very evident Light, that the Authorities cited in my firſt Letter are founded upon the cleareſt Reaſon.

Anaxagoras, you ſay, believed the Sun to be a Lump of red-hot Iron; *Euripides* thought it a Clod of Gold; and others ſtill more ridiculouſly have imagined it to be a dark Body, void of all Heat. That the Sun is a vaſt Body of blazing Matter, notwithſtanding the various Opinions of thoſe primitive Sages, will, I think, hardly admit of a Queſtion: Since the known Warmth of his prolifick Beams, and the viſible Effect of the Burning-glaſs, puts it quite out of the Power of our preſent Set of Senſes, at leaſt to argue againſt it; and how reaſonably we may imagine the Stars to be all of the ſame or like Nature, will ſufficiently appear from theſe following Conſiderations: Firſt, it is well known to all Mathematicians, that any viſible Object of any determined Magnitude may be reduced to the Appearance of * a phyſical Point, by removing the Eye of the Obſerver to a proper or proportionable Diſtance from it, within the finite View: And that the apparent Diameter of every luminous celeſtial Body, will always be diminiſhed reciprocally, in Proportion to the Diſtance from the Eye, till they become altogether imperceptible.

E 2

Thus

* What is here meant by a phyſical Point, is a Point viſible to the naked Eye, which human Art cannot divide; and ſo far it partakes of the Property of a mathematical one, which is only to be conceived, and not ſeen.

Thus the Difk of the Sun, which appears to us at Earth under an Angle of about half a Degree, if feen from the Planet *Saturn*, would appear not much bigger than the Planet *Venus* or *Jupiter*, in their moft neighbouring Vicinity does to us; and confequently to an Eye placed in the Aphelion Point of the Orbit of the great *Comet* of 1680, his apparent Diameter would be fo reduced as to feem but little bigger than the largeft of the Stars; and by the fame Analogy, or Way of Reafoning, admitting Space and Diftance infinite, which I humbly apprehend is not to be difputed, were all the Matter in the Univerfe united, and conglobed in one Mafs, with refpeft to ocular Senfation, it might be diminifhed fo near to a mathematical Punctum, as to be almoft adequate to our Ideas of Nothing.

This to any tolerable Optician, muft be an evident Conviction of the Truth of the modern Aftronomy, which now univerfally allow all thofe radiant Bodies the Stars to be of the fame Nature with the Sun; and that as certainly they are no other than vaft Globes of blazing Matter, all undoubtedly fhining by their own native Light.

But as you have often objected to what has been faid of the Diftance of the Stars in general, and may poffibly from a Suppofition, that they are, or may be, much nearer to us, infer, that their Light, like that of the Planets, may be alfo borrowed from the Sun, or from fome other radiant Body, which, from the Nature of the Suppofition, muft of Confequence be invifible to us, I judge it will not be amifs to throw a few demonftrative Arguments in your Way, in order to lead you a little out of the Path of an early Prejudice, and draw you as it were by Degrees through the Dawn of aftronomical Reafoning, out of your original Error, and refcue your Imagination from the falfe Notions imbibed from Phænomena only in your younger Years. This I guefs cannot fail of reconciling you to this more rational Way of Thinking, and make you acquainted with Truths of much Confequence, which perhaps you have yet been an intire Stranger to. The grand *Deceptio Vifus*, which I muft firft endeavour to remove, and which as a fort of Paradox in Nature, has, as I may fay, imprifoned the Underftanding of many fuperficial Reafoners, and in general all incurious Men, is this.

Moft People are too apt to think originally, that as the Heavens appear to be a vaft concave Hemifphere, that the Stars muft of courfe, as of Confequence, be fixed there, like fo many radiant Studs of Fire, of various Magnitudes; and take it for granted, chiefly defigned for no other Purpofe than to deck and adorn the Canopy of our Night. This was long ago the Opinion of *Thales* the *Milefian*, and wants not the Authority of

many

78

many of the Antients to back it. Others, in particular * *Ptolomy* of *Pelufium* in *Africa*, who from his Experience in this Science, is called by fome the Prince of Aftronomers, believed them to be Loop-holes in the vaft folid celeftial Firmament, emitting the Light of the Cryftalline Heaven through it to all within it. The famous *Diogenes*, Cotemporary with *Plato*, conceived them to be of the Nature of Pumice-ftones, and inclined to an Opinion, that they were the *Spiracula*, or Breathing-holes of Heaven. *Anaxagoras* thought them Stones fnatched up from the Earth by the Rapidity of its Motion, and fet on Fire in the upper Regions above the Moon.

But how ridiculous and abfurd all thefe Opinions and Conjectures really are, will eafily appear, if we but once confider the Nature of an unbounded Æther, and the amazing Property of infinite Space.

This, with what has been faid before, will not a little affift your Imagination towards conceiving the Reafonablenefs of the Notion modern Aftronomers are now confirmed in, of their being abfolutely fo many burning Balls, and which was no doubt, many Years ago, the Opinion of *Manilius*, as is evident from thefe Lines in his Poem of the Sphere.

> For how can we the rifing Stars conceive
> A cafual Production ; or believe
> Of the chang'd Heav'ns the oft renafcent State
> *Sol*'s † frequent Births, and his quotidian Fate.
> SHERBURNE.

And again in the fame Poem:

> The fiery Stars, and Æther that creates
> Infinite Orbs, and others diffipates.
> *Zoroafter*,

---

* *Ptolomy* fuppofed two Heavens above that of the fixed Stars, which he called the eighth ; *viz.* a ninth, the Cryftalline, and a tenth the *Primum Mobile*. See Letter the fecond.

> The facred Sun, above the Waters rais'd,
> Thro' Heav'ns eternal, brazen Portals blaz'd ;
> And wide o'er Earth diffus'd his chearing Ray,
> To Gods and Men to give the golden Day.
> HOMER.

† *Xenophanes* believed the Stars to be no other than Clods fet on Fire, quenched in the Daytime, and rekindled in the Night.

Zoroafter, the firft of all Philofophers we read of who ftudied the Stars, is reported to have believed them of a fiery Nature. *Empedocles* judged them to be Fire æthereal, ftruck forth in its Secretion, and blazing in the upper Regions. *Plato* thought them Fire, with the Mixture of other Elements as Cements. *Heraclides* Worlds by themfelves, of *Earth*, *Air*, and *Fire*; and *Ariftotle*, fimple Bodies of the Subftance of Heaven, but more condenfed.

But that I may not take up too much of your Time with Opinions that has been imbibed in the Infancy of Aftronomy, and has long ago been exploded, I fhall attempt but one Thing more to confirm your Sentiments in this new Doctrine.

Firft, that the Stars are all at a Diftance, not to be determined by the utmoft Perfection of human Art, is manifeft from their having very little, or no fenfible Parallax; and confequently, that any one of them is abfolutely bigger or lefs than another, from the fimple Laws of Opticks, cannot poffibly come under our Obfervation to be afcertained; but that they all of them may be nearly of the fame Size or Solidity, is as impoffible, with any Shew of Reafon to deny, fince it is a known Principle in Geometry, that all vifible Objects naturally diminifh, as has been faid before, or are magnified in a certain Proportion to their Diftance from the Eye; and hence we may conclude, and not without Reafon in its ftrongeft Light to fupport us, that the fmalleft Stars, to the very leaft Denomination, are only removed refpectively more diftant from the Obferver's Station; and that at leaft this we may be certain of, that they are all together undoubtedly an Infinity of like Bodies, diftributed either promifcuoufly, or in fome regular Order throughout the mundane Space: And, as *Marino* fays,

> Refplendent Sparks of the firft Fire!
> In which the Beauty we admire,
> And Light of thofe eternal Rays,
> The uncreated Mind difplays,

It remains now I think to fhew, and endeavour to prove, that the Stars are not only light Bodies of the Nature of the Sun, but that they are really fo many Suns, all performing like Offices of Heat and Gravity, in a regular Order, throughout the vifible Creation, in oppofition to an Opinion

you

* Mr. *Bradley*, Aftronomer Royal, has, in a great meafure, proved that the Aberration of the Stars hitherto miftaken for a Parallax, may arife from, and indeed feems to be no other than the progreffive Motion of Light, and Change of Place to the Eye, arifing from the Earth's annual Motion and Direction.

80

you have formerly hinted at, of their being in another Senfe of a fecondary Nature.

All Objects within the fenfible Sphere of the Sun's Attraction, or Activity, are in fome meafure magnified by a good Telefcope: But the Stars are all placed fo far without it, that the beft Glaffes has no other Effect upon them than making them appear more vivid or lively, but all inate opaque Bodies, reflecting only a borrowed Light from fome primary one, contrary to this Property, are all obferved to lofe their Light, in the fame Proportion, as they are magnified, and through all Glaffes become more dull than otherwife they appear to the naked Eye: And hence we may infer, without any further Evidence, that the Stars are all light Bodies endowed with native Luftre; and that Bodies, like the known Planets, from the fame Reafoning, it is as clear they cannot be, becaufe their Diftance, though uncertain as to the Truth of the whole, yet fuch a Part of it as cannot be denied, would render them all in fuch a Cafe invifible.

A Proof of this will plainly prefent itfelf, if we confider the Courfe of the known Comets, who all of them, without Exception, become imperceptible, and intirely difappear; though moft of them much bigger than the Earth, or any of the leffer Planets, long before they arrive at their refpective Aphelions.

But we are under a kind of Neceffity to believe them either Suns or Planets, that is either dark or light Bodies; and fince I have fhewn the Improbability; nay, I may venture to fay, the Impoffibility of their being the firft, it is natural fure to conclude, that they muft be of the laft Sort; and I am perfuaded, if you but once confider how ridiculous it is to imagine fo vaft a Number of Bodies, all rolling round a Number of invifible Suns, which muft otherwife be the Cafe, fince they are feen on all Sides of ours, and cannot poffibly be enlightened by him, or any, how all of them, by any one elfe, you cannot poffibly have any fort of Difficulty in this Determination: But that no Arguments may be wanting to enforce your Belief of what is here concluded, it will not be amifs to put you in Mind of an optical Experiment or two, which cannot fail of convincing you of the vaft Probability of what is here afferted of them; and next to a moral Certainty, demonftrate the Truth of what fo many of the beft Aftronomers have advanced, as before namely, that the Stars are all, or moft of them, Suns like ours.

Place any concave Lenfe before your Eye, and you will find all vifible Objects will appear through it, as removed to a much greater Diftance than they really are at, and reciprocally as much diminifhed. Now, if
you

81

you look upon one of these Glasses of a proper Concavity, opposed to the Sun or Moon, you will respectively have the Appearance of a real Star. or Planet, the first exhibited by the Body of the Sun, the other by the Moon, and either more or less diminished in Proportion to the Surface of the Sphere the Glass is ground to.

For Example, a double Concave, or Glass of a negative Focus, ground to a Sphere of about three Inches Diameter, will if opposed to the Sun's Disk at a proper Distance from the Eye, help you to a very good Idea how the Sun appears to the Planet *Jupiter*; and if a proper Regard be had to the Distance of the Planet *Saturn*, a Lense still more concave may be formed to give a just Idea of the Sun's Appearance to *Saturn*. Again, one much more concave than the former, proportioned to the Orbit of *Mars*, will naturally exhibit the solar Body, as seen from that Planet.

To the Planet *Venus* and *Mercury*, the Sun appearing much larger than to us at the Earth, to have any tolerable Notion of his varied Phænomena to them, it will be necessary to procure Glasses of a suitable Convexity, ground to reciprocal Concaves, which may easily be done to any Focus, so as to shew how the Sun, naturally appears to the Inhabitants of those two Planets.

The various Appearances of the Planets themselves to us at the Earth, may also well enough be had, if through Glasses analagous to their respective Distance and Magnitude, we look at the Moon, particularly all the Phases of *Venus*, and even of *Mercury*, and the Gibosity of *Mars*, &c. may be justly and beautifully represented at different Ages of the Moon, as those Planets appear through the largest and best Telescopes.

This Way you may convince even your Friend * * *, who you tell me has reasoned all his Senses useless, and yet continues so great an Atheist in Astronomy, as not to believe the World turns round upon its Axis, though he gives no better Reason for it than that of his not being giddy.

After all these Arguments, I hope no new Difficulties will arise to retard your Belief, or deprive the Stars of their solar Nature, so justly due to them: This Point gained, the next Thing to be considered is, whether all those glorious Bodies, the far greater Part of whom being invisible to the naked Eye, were made purely and purposely for the sole Use of this diminitive World, our little trifling Earth.

> ——Men, conceited Lords of all,
> Walk proudly o'er this pendent Ball,
> Fond of their little Spot below,
> Nor greater Beings care to know,
> *But think those Worlds, which deck the Skies,*
> *Were only form'd to please their Eyes.*   DUCK.

The

The very Suppofition not only implies a profound Ignorance of the Divine Attributes, but is as impious, and full of Vanity, as it is erroneous and abfurd, and even a Blindnefs fufficient of itfelf, were there no other Caufe for it, to introduce Idolatry in the Minds of Mortals, by finking the divine Nature fo near to the human.

It being granted that the Stars are all of the fame Kind, I think it may be agreed, that what we evince of any one may be allowed to be true of any other, and confequently of all the reft. This *Poftulata* gained, I fhall next proceed to enquire what the real Ufe and Defign of fo many radiant Bodies are, or may be made for.

The Sun we have juftly reduced to the State of a Star, why then in Reafon fhould he have his attendant Planets round him, more than any of the reft, his undoubted Equals? No Shadow even of a Reafon can be given for fuch an Abfurdity.

May we not with the greateft Confidence imagine, that Nature as juftly abhors a *Vacuum* in Place, as much as Virtue does in Time? Surely yes: And by fuppofing the Infinity of Stars, all centers to as many Syftems of innumerable Worlds, all alike unknown to us; how naturally do we open to ourfelves a vaft Field of Probation, and an endlefs Scene of Hope to ground our Expectation of an *ever*-future Happinefs upon, fuitable to the native Dignity of the awful Mind, which made and comprehends it; and whofe Works are all as the Bufinefs of an Eternity?

If the Stars were ordained merely for the Ufe of us, why fo much Extravagance and Oftentation in their Number, Nature, and Make? For a much lefs Quantity, and fmaller Bodies, placed nearer to us, would every Way anfwer the vain End we put them to; and befides, in all Things elfe, Nature is moft frugal, and takes the neareft Way, through all her Works, to operate and effect the Will of God. It fcarce can be reckoned more irrational, to fuppofe Animals with Eyes, deftined to live in eternal Darknefs, or without Eyes to live in perpetual Day, than to imagine Space illuminated, where there is nothing to be acted upon, or brought to Light; therefore we may juftly fuppofe, that fo many radiant Bodies were not created barely to enlighten an infinite Void, but to make their much more numerous Attendants vifible; and inftead of difcovering a vaft unbounded defolate Negation of Beings, difplay an infinite fhapelefs Univerfe, crowded with Myriads of glorious Worlds, all varioufly revolving round them; and which form an Atom, to an indefinite Creation, with an inconceivable Variety of Beings and States, animate and fill the endlefs Orb of Immenfity.

F

That

That the sidereal Planets are not visible to us, can be no Objection to their actual Existence, and being there, is plain from this; it is well known, that the Stars themselves, which are their Centeral, and only radiant Bodies, are little more to us at the Earth, than mathematical Points. How ridiculous then is it to expect, that any of their small opaque Attendance, should ever be perceived so far as the Earth by us; and besides, to show the Impossibility of such a Discovery, we need only consider, what is, and what is not to be expected, or known in our own home System. All the Planets in this our sensible Region, every Astronomer knows, is far from being visible to one another, in every individual Sphere; for to an Eye at the Orb of *Saturn*, this Earth we live upon, which requires Years to circumscribe, and Ages to be made acquainted with, and is far from being yet all known, cannot possibly from the above Planet be seen: And further, since *Saturn* and *Jupiter*, two of the most material and considerable Globes we know of, except the Sun himself, are Bodies apparently of the same kind, and are observed to have each a Number of lesser Planets moving round them; why may we not expect with equal Certainty and Propriety, that all other Bodies, under the same Circumstances, are in like manner attended; that is, seeing the Sun is found to be the Center of a System of Bodies, all variously volving round him? where lies the Improbability of his fellow Luminaries, the Stars, being surrounded in like sort, with more or less of such Attendance.

I shall offer but one Thing more to your Consideration in this Affair, and which I am in great Hopes will be sufficient to make you think these natural Suggestions a good deal more than probable, and that is this:

The modern Astronomers having, in a great measure, proved that the Stars are, in all respects, vast Globes of Fire like our Sun. Let us suppose a new-created Mind, or thinking Being, in a profound State of Ignorance, with regard to the Nature of all external Objects, but fully endowed with every human Sense and Force of Reason, suspended in Æther, exactly in the midway, betwixt * *Syrius* and the Sun; in which Case, both of these Luminaries would equally appear much about the Brightness of the largest of our Planets. Now should such a Being, determined either by Accident or Choice, arrive at this our System of the Sun, and seeing all the planetary Bodies moving round him, I would ask you what you think he would imagine to be round *Syrius*? Your Answer, I think I may venture to say, would not be *nothing*; and methinks I already hear you say, Why Planets such as ours.

* A Star of the first Magnitude in the greater-*Dog*, and the most neighbouring to our Sun.

PLATE

The SUN

PLATE XI.

THE

S

SUN

ORBIT of MERCURI.

P

SYRIUS

P

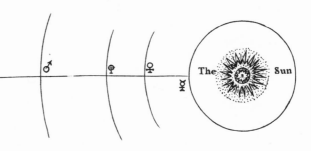

The ☉ Sun

PLATE XI.

Is defigned as a geometrical Scale to all the primary Parts of the vifible Creation, with regard to the Diftance of Orbits compared with the Globe of the Sun; by which at once may be conceived, and juftly meafured in the Mind, not only the mean Diftance of the Planets with regard to one another, but alfo that of the Comets, and even the comparative Diftances of the neareft of the Stars, which will, I guefs, greatly help you to form an Idea of the vaft Extent of Space neceffary to comprehend the whole Creation.

Fig. 1. Is a Radius of the Orbit of *Mercury*, in true Proportion to the Body of the Sun reprefented at S, fhewing at the fame time a fmall Portion of the opaque Planet's Orbit, and the real Length of its Shadow at P.

Fig. 2. Is a Radius of the whole Syftem of the Planets as far as the Orbit of *Saturn* in Proportion to a compleat Orbit of *Mercury*, much lefs than the former; the former ferving as a better known Scale to confider the amazing Diftances of the more remote Planets by.

Laftly, *Fig.* 3. Is a Reprefentation of the leaft poffible Diftance of *Syrius* and the Sun, proportionable to the Magnitude of the Sphere of our Comets, *&c.* reprefented at S, whereby it evidently appears, that as all the Planets of *Syrius* muft be included within the fmall Sphere reprefented in the Center P, none of them could poffibly be feen at the Sun, not only by reafon of the Smallnefs of the Angle of Suftenfion, or Elongation, but alfo as being loft in the fuperior Light of *Syrius* himfelf, in fo minute an Orb of Vicinity.

Confequently (as you muft perceive) no Arguments can poffibly be drawn to deny the Exiftence of fuch Bodies, with any Shew of Reafon, from their not having been feen by us.

Here I muft obferve to you, that you cannot confider this Scale of Orbits too much before you look upon Plate XVII.

To conclude, it evidently feems to be the End and Defign of Providence, by this vifible Variety of Beings, to lift the Minds of Men above this narrow Earth, in Search of that powerful Being upon which we are all fo much dependant; and the *Creator*, no doubt, in this vaft Difplay of his Wifdom and Power, defigned the amazing Whole, as the adequate Object of every Part, and as fuch equally open on all Sides, to the penetrating Progrefs of human Minds, and through the moft extenfive Faculty of Senfe, the *Sight*, to draw our Reafon and Underftanding by Degrees, from finite Objects into Infinity; and as the laft Refult of celeftial Contemplations place within our Reach, a certain Evidence of a future State, *and the manifeft Manfions of Rewards and Punifhments, fuited no doubt moft equitably to all Degrees of Virtue, and to every Vice.*

When

" When I confider (fays Mr. *Addifon*, fpeaking as having taken particular
" notice of a fine Evening) that infinite Hoft of Stars, or to fpeak more
" philofophically of Suns, which were then fhining upon me, with thofe
" innumerable Sets of Planets or Worlds, which were then moving round
" their refpective Suns; when I ftill enlarge the Idea, and fuppofed ano-
" ther Heaven of Suns and Worlds rifing ftill above this which we dif-
" covered; and thefe ftill enlightened by a fuperior Firmament of Lu-
" minaries, which are planted at fo great a Diftance, that they may ap-
" pear to the Inhabitants of the former as the Stars do to us; in fhort,
" whilft I purfued this Thought, I could not but reflect on that little
" infignificant Figure which I myfelf bore amongft the Immenfity of
" God's Works:" This Reflection, I judge, as you are an Admirer of the
Author, you will not look upon as impertinent in this Place, efpecially as
it muft enforce what I have endeavoured to fhew you, namely, the Reafonablenefs of a Plurality of fidereal Syftems, and their Multiplicity of
Worlds; which, if you are yet in Doubt of, I hope you will at leaft for-
give fo well defigned an Attempt with your ufual Candour.

I am now prepared to proceed in the chief Defign of this Undertaking,
which is to folve the Phænomena of the *Via Lactea*; and propofe in my
next to anfwer more fully your farther Requeft.

I am, &c.

L E T T E R

LETTER the FIFTH.

Of the Order, Diftance, and Multiplicity of the Stars, the Via Lactea, *and Extent of the vifible Creation.*

S I R,

WE are told, and, if I remember right, it is alfo your Opinion, that three of the fineft Sights in Nature, are a rifing Sun at Sea, a verdant Landskip with a Rainbow, and a clear Star-light Evening : All of which I have myfelf often obferved with vaft Delight and Pleafure. The firft I have frequently beheld, and always with an agreeable Surprize ; the fecond I have as often taken notice of, with no fmall Degree of Admiration ; but the laft I fhall never look up to without an Aftonifhment, even mixed with a kind of Rapture. The Night you laft left us, this admirable Scene was in its full Beauty ; and, as *Milton* fays,

> Silence was pleas'd : now glow'd the Firmament
> With living Saphirs ; *Hefperus* that led
> The ftarry Hoft rode brighteft.——

I found it was impoffible to look long upon this ftupendious Scene, fo full of amazing Objects, and particularly the *Via Lactea*, which (the Moon being abfent) was then in great Perfection, without being put in Mind of my Tafk. This furprizing Zone of Light being the chief Object I have undertaken to treat of and demonftrate.

This amazing Phænomenon which have been the Occafion of fo many *Fables*, idle Romances, and ridiculous Opinions amongft the Antients, ftill continues to be unaccounted for, and even in an Age vain enough to boaft Aftronomy in its utmoft Perfection.

What will you fay, if I tell you, it is my Belief we are fo far from the real Summit of the Science, that we fcarce yet know the Rudiments of what may be expected from it. This luminous Circle has often engroffed my Thoughts, and of late has taken up all my idle Hours ; and I am now in

great

great Hopes I have not only at laſt found out the real Cauſe of it, but alſo by the ſame Hypotheſis, which ſolves this Appearance, ſhall be able to de-monſtrate a much more rational Theory of the Creation than hitherto has been any where advanced, and at the ſame Time give you an intire new Idea of the Univerſe, or infinite Syſtem of Things. This moſt ſurprizing Zone of Light, which have employed ſucceſſively for many Ages paſt, the wiſeſt Heads amongſt the Antients, to no other Purpoſe than barely to deſcribe it; we find to be a perfect Circle, and nearly biſecting the ce-leſtial Sphere, but very irregular in Breadth and Brightneſs, and in many Places divided into double Streams.

    * The principal Part of it runs through the *Eagle*, the *Swan*, *Caſſiopea*, *Perſeus*, and *Auriga*, and continues its Courſe by the Head of *Monoceros*, along by the greater *Dog* through the Ship, and underneath the *Centaur*'s *Feet*, till having paſſed the *Alter*, the *Scorpion's Tail*, and the Bow of *Aquarius*, it ends at laſt where it begun.

PLATE XII, and XIII.

Repreſents the two Hemiſpheres, where its true Tract is diſtinguiſhed amongſt the principal Stars, and may eaſily be conceived by them to cir-cumſcribe and biſect the whole Heavens.

This is that Phænomena I am about to explain and account for; but before I proceed farther, I judge it will be no *improper Precognita*, to give you the Thoughts of the Antients upon it; the Relation perhaps may re-quire ſome Patience; but I gueſs, that after reading ſuch wild and extra-vagant Notions concerning it, you will naturally judge more favourably of the Conjectures of the Moderns upon it, and particularly of what is con-cluded in the ſucceeding Pages.

Theophraſtus

    * ————— Carried toward the oppoſed *Bears*,
Its Courſe cloſe by the *Artick* Circle ſteers,
And by inverted *Caſſiopea* tends;
Thence by the *Swan* obliquely it deſcends
The Summer Tropick, and *Jove's* Bird divides;
Then croſs the Equator, and the Zodiack glides
'Twixt *Scorpio's* burning Tail, and the left Part
Of *Sagitarius*, near the fiery Dart;
Then by the other *Centaur's* Legs and Feet,
Winding remounts the Skies (again to meet)
By *Argos'* Topſail, and Heav'ns middle Sphere,
Paſſing the *Twins*, t' o'ertake the Charioteer;
Thence *Caſſiopea* ſeeking thee does run,
O're *Perſeus* Head, and Ends where it begun.

SHER. MANILIUS.

PLATE XII.

PLATE XIII

Theophraftus * was of Opinion, that the Hemifpheres, which, by many of the Antients were imagined to be folid, was joined together here ; and that this was the foldering of the two Parts into one. † DIODORUS thought it celeftial Fire, of a denfe and compact Nature, feen through the Clifts or Cracks of the parting Hemifphere : But as *Manilius* fays,

> Aftonifhment muft fure their Senfes reach,
> To fee the World's wide Wound, and Heav'n's eternal Breach.

OENOPIDES ‖ believed it the ancient Way of the Sun, till frighted at the bloody Banquet of *Thyeftis*.   ** ERATOSTHENES fuppofed it *Juno's* Milk, fpilt whilft giving Suck to *Hercules*. ‡‡ PLUTARCH makes it the Effect of *Phaeton's* confufed Erratication ; but I think it is plain ‡‡ OVID judged them to be Stars, and the ancient *Ethnicks* believed them to be the blifsful Seats of valiant and heroic Souls.

> ——Valiant Souls, freed from corporeal Gives,
> Thither repair, and lead æthereal Lives.
>   MANILIUS.

* *Macrobius*, lib. i. cap. 15.
> Or meets Heaven here ! and this white Cloud appears
> The Cement of the clofe-wedg'd Hemifpheres!

† The facred Caufes human Breafts enquire,
> Whether the heavenly Segments there retire,
> (The whole Mafs fhrinking, and the parting Fame
> Thro' cleaving Chinks admits the ftranger Flame.

‖ Or feems that old Opinion of more Sway,
> That the Sun's Horfes here once run aftray,
> And a new Path mark'd in their ftraggling Flight,
> Of fcorching Skies, and Stars adufted Light.

** Nor muft that gentle Rumour be fuppreft,
> How Milk once flowing from fair *Juno's* Breaft
> Stain'd the celeftial Pavement, from whence came
> This milky Path, its Caufe fhewn in its Name.

‡‡ When from the hurried Chariot Light'ning fled,
> And fcatter'd blazes all the Skies o'erfpread ;
> By whofe Approach new Stars enkindled were,
> Which ftill as Marks of that fad Chance appear.

>   MANILIUS.

‡‡ A Way there is in Heaven's expanded Plain
> Which when the Skies are clear, is feen below,
> And Mortals by the Name of *Milky*, know,
> The Ground-work is of Stars -----

>   *Ovid's* Met. lib. i.
>   But

But * Democritus long ago believed them to be an infinite Number of small Stars; and such of late Years they have been discovered to be, first by *Gallaleo*, next by *Keplar*, and now confirmed by all modern Astronomers, who have ever had an Opportunity of seeing them through a good Telescope.

PLATE XIV.

Is from an Observation I made myself, of a bright Part of this Zone near the Feet of *Antinous*; which, (by a Mistake of the Engraver) is, as it appears through a Tube of two convex Glasses. I saw it through a very good Reflector, and formed the Plan by a Combination of Triangles.

Milton takes notice of this Zone in a most beautiful Manner, where he describes the Creator's Return from his six Day's Work to Heaven, he introduces it as a Simile to express his Idea of the eternal Way, or Road to the celestial Mansions.

> —— A broad and ample Road, whose Dust is Gold
> And Pavement Stars, as Stars to thee appear,
> Seen in the *Galaxie*, that Milky Way,
> Which nightly as a circling Zone thou seest
> Powder'd with Stars.

But to infer from their Appearance only, that they are really Stars, without considering their Nature and Distance; and that nothing but Stars could possibly produce such an Effect, may perhaps be assuming too much, when we have nothing but the bare Credit of the *Belgic* Glasses to support our Conjectures; and although this may be sufficient for any Mathematician, yet for your greater Satisfaction, I have thought proper to give two or three more evincing Arguments, to confirm these important Discoveries. *Democritus*, as I have said before, believed them to be Stars long before Astronomy reaped any Benefit from the improved Sciences of Optics; and saw, as we may say, through the Eye of Reason, full as far into Infinity as the most able Astronomers in more advantageous Times have done since, even assisted with their best Glasses: And his Conjectures are almost as old as the philolaic System of the Planets itself; the Construction of which, though attempted by many, none have ever yet been able to confute.

The Light which naturally flows from this Crowd of radiant Bodies is mixt and confused, chiefly occasioned by the Agitation of our Atmosphere, and from a Union of their Rays of Light, by a too near Proximity of their Beams, altogether they appear like a River of Milk, but more of a pelucid Nature, running all round the starry Regions.

For

* *Plutarch (in Placitis Philosoph.)*

PLATE XIV.

For in the azure Skies its candid Way
Shines like the dawning Morn, or cloſing Day.

There are alſo many more ſuch luminous Spaces to be found in the Heavens of the ſame Nature with theſe, which we know to be Stars; in particular the *Nebulæ*, or cloudy Star in the *Præſepe* of 36; a cloudy Star in *Orion* of 21; * a cloudy † Knot not far from this in the ſame Aſteriſm of 80; in one Degree of the ſame Conſtellation 500, and in the whole Form above ‡ 2000. All of which are great Confirmations of the Truth of our Aſſertion, *i. e.* that this Zone of Light proceeds from an infinite Number of ſmall Stars. Here it will not be amiſs to obſerve, that it has been conjectured, and is ſtrongly ſuſpected, that a proper Number of Rays, meeting from different Directions, become Flame; and that hence it may prove not the Sun's real Body which we daily ſee, but only his inflamed Atmoſphere. I begin to be of Opinion, and I think not without Reaſon, that the true Magnitude of the Sun is not near what the modern Aſtronomers have made it; and that it may not poſſibly be much above two Thirds of what it appears to us; I don't mean that this Expanſion of the ſolar Flame is any Part of that dilated Light mentioned by Sir *Iſaac Newton*, and conceived to be round all light Bodies in general; but you may conſider it as not much differing from it, not of an unlike Nature, only greater in Degree, and peculiar to the Sun and Stars, who are all, as has been before in a manner demonſtrated to be actually Globes of Fire.

This, tho' I preſume to call it at preſent only meer Hypotheſis, will in a great meaſure account for the exceſſive Changes in the Conſtitution of our Air and Atmoſphere, which we often find very unnatural to the Seaſon; alſo be a Means perhaps of reconciling the vaſt Diſproportion ſo very remarkable betwixt the Sun and the leſſer Planets, and many other Circumſtances in the Syſtem of no ſmall Conſequence in Aſtronomy: One of which Particulars you have frequently expreſſed a great Miſtruſt and Diſapprobation of, as ſuſpecting ſome kind of a Fallacy in the Computation; and the other is Matter of general Complaint, being by many attributed to a Change in the Direction of the Earth's Axis ‖; and by ſome, eſpecially the Vulgar, to too near an Approximation of the Earth to ſome one of the celeſtial Bodies. But all this will very naturally be accounted for by the Levity, or expanding Quality of the Sun's circumambient

---

* Vide *Galilæo.*
† Betwixt the Sword and Girdle of *Orion.*
‡ Vide *Reitha.*
‖ Which, through Ignorance of the true Caſe, is commonly called a Shock, a Bruſh, or Shove.

G

Flame,

Flame, or Atmofphere; and hence, according to its various State, being more condenfed, or rare, we may have Heat or Cold in the greateft Extream, and alternately fo, in a perpetual Viciffitude.

The Truth of this Doctrine will evidently appear from the Obfervations of the Sun's Diameter through the Year 1660, by the indefatigable *Mouton:* And, I muft own, I am not a little furprized to find that no Conclufions have been drawn from them of this Kind. I am perfwaded, if you once compare thofe Numbers, you will be very far from thinking this an improbable Suggeftion. But this Digreffion has led me a little too far from the *Via Lactea,* and too near home again ; I muft now think of returning to the Stars, and my next Endeavours muft be to give you fome Idea of the Number of them. Through very good Telefcopes there have been difcovered in many Parts of this enlightened Space, and even out of it, feveral thoufand Stars in the Compafs of one fquare Degree ; in particular near the Sword of *Perfeus,* and in the Conftellations of * *Taurus* and *Orion.*

PLATE XV.

Reprefents the *Pleides,* a well known Knot of Stars in the Sign *Taurus,* as they appeared to me thro' a one Foot reflecting Telefcope : And *Plate* XVI. is a View of the *Perfides,* another furprizing Knot of Stars in the Conftellation *Perfeus,* exactly as they appear through a Tube of two convex Glaffes. There are alfo other luminous Spaces in the ftarry Regions, not unlike the Milky Way, which I have had no Opportunity of obferving ; fuch as the *Nebeculae,* near the South Pole, called by the Seamen *Magellanic* Clouds ; and which likewife viewed through Telefcopes, prefent us with little *Nebulae,* and fmall Stars interperfed : One of thefe Kind is fituated between *Hydrus* and *Dorado* ; and another, fomething lefs than this, betwixt *Hydrus* and the *Toucan.*

Now admitting the Breadth of the *Via Lactea* to be at a Mean but nine Degrees, and fuppofing only twelve hundred Stars in every fquare Degree, there will be nearly in the whole orbicular Area 3,888,000 Stars, and all thefe in a very minute Portion of the great Expanfe of Heaven. What! a vaft Idea of endlefs Beings muft this produce and generate in our Minds; and when we confider them all as flaming Suns, Progenitors, and *Primum Mobiles* of a ftill much greater Number of peopled Worlds, what lefs than an Infinity can circumfcribe them, lefs than an Eternity comprehend them,

---

* *Galilaeo* in one cloudy Star of this Conftellation, difcovered no lefs than twenty-one, and in that of the *Praefepe* thirty-fix.

or

PLATE XVI.

or lefs than Omnipotence produce and fupport them, and where can our Wonder ceafe?

In this Place perhaps I ought not to pafs over the aftonifhing Pheno-menon of feveral new Stars, &c. which have frequently appeared, and foon again vanifhed, in the fame Point of the Heavens. But as the Bu-finefs of this Theory is rather to folve the general, than any particula Phænomenon, I fhall only here by way of Note fubjoin a Table of fuch as has been regularly obferved, and by whom they were firft difcovered.

A Table of feveral new Stars, Nebulæ, and double Stars, &c.

Nomina Stellarum.	Obfervationum.
Septima Pleiadum	Loft after the burning of *Troy*, but now returned; fee *Ricciolus*.
A new Star appeared in *Caffiopea*, nearly in the fame Place with that of 1572.	*Anno Dom.* 945, bright as *Jupiter*; fee *Ricciolus*.
The new Star in *Caffiopea*'s Chair.	Bright as *Venus*, from *November* 1572 to *March* 1574.
A new Star in *Collo Ceti*.	Of the 3d Magnitude, is faid to have appear'd periodically, feven Times in fix Years, *i. e.* every three hundred and thirteen Days: It was firft obferved in *Auguft* 1596, for two Months, by D. *Fabricius*
A new Star in the Swan's Neck,	Obferved by *Kepler* in 1600, of the third Magnitude, till the Year 1659; then gradually decreafing; in 1661 it difappeared; in 1666 it became vifible again, and is yet to be feen of the fixth Magnitude.
A new Star in the Right Foot of *Serpen-tarius*,	Bright as *Venus* from *October* 1604 to *October* 1605: fee *Kepler*.
A new Star in *Andromeda*'s Girdle,	Seen by *Simon Marius* and *Fabricius, Anno* 1612.
A new Star in *Antinous*,	Seen by *Juftus Byrgius*.
A new Star feen in the Whale,	In 1638, by *John Procyclides Holuarda*, of the third Mag-nitude, which difappeared periodically, every three hundred and thirty Days.
A new Star in the Fox's Head,	Of the third Magnitude, feen by *Hevelius* in *July* 1670, and till *Auguft* 1671, alfo from *March* 1672 to *Septem-ber* 1672.
A new Star in the Swan's Neck.	This appear'd periodically every four hundred and four Days, and about fix Months at a Time; it was feen at its brighteft, *September* 10, 1714.

Of the Nebulæ, *or Cloudy Stars.*

Nebulofe in *Orion*'s Sword.	
Nebulofe in *Andromeda*'s Girdle.	Small, but very luminous.
Nebulofe in the Bow of *Sagitarius*,	Never feen in *England*.
Nebulofe in *Centaurus*,	
A *Nebulofe* preceding the right Foot of *Antinous*,	Obfcure, but with a Star in the Middle of it.
Nebulæ in *Dorfo Herculis*,	Difcovered by Dr. *Holly*.

Befides the *Nebulæ*, and new Stars, it appears from the ancient Catalogues of *Hevelius*, &c. that fome of the old ones have intirely vanifhed; in particular, one in the left Thigh of *Aquarius*, the contiguous one preceding in the Tail of *Capricorn*; the fecond on the Belly of the Whale; the firft of the unformed ones after the Scales of *Libra*, and feveral others. Many of the Stars alfo appear to be double, as the firft Star of *Aries* and *Cafter*; others triple, as one in the *Pleiades*; and the middle one in *Orion*'s Sabre; and others again, quadruple, &c.

G 2   I would

I would now willingly help you to conceive the indefinite mutual Diſtance of the Stars, in order to give you ſome ſmall Notion of the Immenſity of Space; but as this will be a Taſk merely conjectural, I ſhall only deſire you to believe it as far as your Reaſon will carry you, ſafely ſupported by an obvious Probability.

Perhaps it may be neceſſary here to acquaint you, that all the Stars are ſo far apparently of different Magnitudes, that no two of them are to be found in the whole Heavens exactly the ſame, either in Bigneſs or Brightneſs *. The largeſt we have ſufficient Reaſon to believe is the neareſt to us; the next in Bigneſs and Brightneſs more remote; and ſo on to the leaſt we ſee, which we judge to be the moſt remote of all.

The firſt Degree, or that of the largeſt Magnitude, we give to Syrius, the ſecond to Arcturus, the third to Aldebaran, the fourth to Lyra, the fifth to Capella, the ſixth to Regulus, the ſeventh to Rigel, the eighth to Fomahaunt, and the ninth to Antarus: Theſe are all ſaid to be of the firſt Claſs; and beſides which, there are at leaſt, within the Reach of our lateſt improved Opticks, nine more Denominations within the Radius of the viſible Creation.

Now, by the certain Return of the Comets, which we find are all governed by the Laws of this Syſtem, and ſuppoſed to be undiſturbed by any of the others, we cannot avoid concluding, if we conſider them at all to the Purpoſe, that the neareſt Stars cannot be leſs diſtant than twice the Radius of the greateſt Orbit belonging to the Sun. Moſt Mathematicians think this a great deal too near, as it muſt of courſe make all the Syſtems join, as in Contact; and I think we may ſafely add, to ſeparate their Spheres of Attraction, at leaſt one Half of this Diſtance more, which will make in the Whole about four hundred and twenty Semi-orbits of the Earth, or 33,600,000,000 Miles. This even the ingenious Mr. *Huygins* endeavours to prove ſtill much too little, and his Arguments are ſuch as cannot eaſily be refuted. His Principle is grounded upon the known Laws of Analogy, as conſidered in the Proportion of light Surfaces, and is as follows. Having reduced the Sun's Diſk to the Appearance of the Star Syrius, by the Help of a ſmall Hole at the End of his Teleſcope, and comparing this Part of his Surface to the whole Diſk of the Sun, he infers that the Stars Diſtance to that of the Sun muſt be as 27,664 to 1. Hence *Syrius* from us will be nearly (avoiding Units) 2,213,120,000,000 Miles: But this I take to be as much too large as the former is too little; yet, as

---

* A very little Knowledge in Opticks will render this indiſputable, and has been in a great meaſure demonſtrated before; 1. in the Great Dog; 2. in Bootes; 3. in the Bull; 4. in the Harp of *Apollo*; 5. in *Auriga*; 6. in the Lion; 7. in *Orion*; 8. in the Southern Fiſh; 9. at the End of *Erridanus*.

         Mr.

Mr. *Bradley* has, with some Shew of Reason, banished all the Stars out of the Sphere of Parallax, the last is the only Method we can possibly make use of with any kind of Confidence; and Sir *Isaac Newton* endeavours to recommend it with great Force of Argument, as the only probable Means by which we can give any tolerable Guess at these immense Measurements of Space.

To moderate the Matter then if you please, allow me but to make use of a Mean betwixt the two fore-mentioned Numbers; and we may take it for granted, a Distance sufficiently exact, to suit all our Wants in the present Case, namely, to give a very tolerable Idea of the Extent of the visible Creation, which is all I propose in this Place to attempt; but I mean to be much more exact in another.

Now as the Distance from the Sun to the Earth is so small in Proportion to the Distance of the Stars from us, and from one another, we may very well consider the Sun as the Center of our Station, or Position in the general System or Frame of Nature. And as the Stars are very visible thro' good Telescopes, to the ninth or tenth Magnitude, if we multiply the primary Distance of *Syrius*, or of any other of his Class, by this Number of common intermediate Spaces, the Product will be equal to the Radius of the visible Creation to the solar Eye; which, by this Rule, you will find in capital Numbers to be * nearly 6,000,000,000,000 Miles, taking in a Star of the sixth Magnitude, and to a Star of the ninth, 9,000,000,000,000 Miles: But this Computation supposes a mean common Distance of the Stars in a sort of Syzygia, or Direction of a Right Line, which is not the real Case; for the Stars cannot be supposed to diminish in a proportional Magnitude by any mathematical *Ratio*, but by some geometrical, or rather musical one; for Instance, if the Distance of a first be 3, that of a second should be about 5, and of a proportional Third 8,333, *&c. ad infinitum*: But as their true proportional Distance is unknown, the above will be sufficient for our present Purpose; which is only to shew, without Exaggeration, the Space we now are truly sensible of.

This I have here considered more extensively, to obviate all Objections that you may make to the Probability of the general Motion of the Stars, by shewing no Difficulty can possibly arise from their apparent Proximity, Number, or irregular Distribution: Their Distances being so immensely large, no Disorder or Confusion can be supposed in any Direction of them, or Motion whatever. The greatest Distance of the Planets, which all move undisturbed round the Sun, is about three hundred and fifty-three Million of Miles: But the least Distance of one Star from another, is

---

* If the Distance of the Sun and Earth is found too much, which I must own I have a violent Suspicion of, these Numbers must be reduced in like Proportion.

upwards

upwards of two thoufand eight hundred and thirty-two Times that Diftance, or one Million of Millions of Miles: And as no fenfible Diforder can be obferved amongft the folar Planets, what Reafon have we to fuppofe any can be occafioned amongft the Stars, or that a general Motion of thefe primary Luminaries round a common Center, fhould be any way irrational, or unnatural?

What an amazing Scene does this difplay to us! what inconceivable Vaftnefs and Magnificence of Power does fuch a Frame unfold! Suns crowding upon Suns, to our weak Senfe, indefinitely diftant from each other; and Miriads of Miriads of Manfions, like our own, peopling Infinity, all fubject to the fame Creator's Will; a Univerfe of Worlds, all deck'd with Mountains, Lakes, and Seas, Herbs, Animals, and Rivers, Rocks, Caves, and Trees; and all the Produce of indulgent Wifdom, to chear Infinity with endlefs Beings, to whom his Omnipotence may give a variegated eternal Life.

The aftonifhing Diftance of the ftarry Manfions undoubtedly was defign'd to anfwer fome wife End: One Confequence is this, and probably is not without its Ufe: To every Planet of the fame Syftem, the fame fidereal Face of Heaven appears without the leaft Degree of Change; and as the remoteft Regions upon Earth fee the fame Moon and Planets, fo alfo the Inhabitants of the moft diftant Planets in ours, or in any other Syftem, fee the fame Forms and Order of the Stars in common with the reft. The whole Sphere of Heaven being common and unchangeable through all their various Revolutions.

Thus thofe (the People) in the Planet *Venus* will fee the Conftellation of *Orion* juft as we do, and the People in the Planet *Saturn*, much farther ftill removed, alike will view this Conftellation in all refpects the fame; here then, (in the Syftem of the Sun) the Eye removed from us muft only hope to find a new Earth furrounded with the fame fort of Sky: But Beings in another Syftem, behold not only a new Heaven above, but alfo new Earths below; and all the Frame of Nature to them puts on a new Drefs, new Signs, new Seafons, and new Planets roll, and a new Sun renews the Day.

The Heathen Fables here are all erafed with all the Immortality of their vain earthly Gods and Heroes; *Perfeus* and *Alcides* are no more, and both the *Bears* are vanifhed; the *Pleiads* and the *Hyads* join, and fhining *Leo*, though boafting two Stars of the firft Magnitude with us, there no where can be found, loft in the common undiftinguifhed Herd. But ftill Aftronomy will exift, and new-framed Forms may fill the varied Scene.

Perhaps you may expect that I fhould here give you my Conjectures of what fort of Beings may be fuppofed to refide in the *Ens Primum*, or *Sedes Beatorum*

Beatorum of the known Univerfe, whether mortal, immortal, or Creatures partaking in fome Degree of the Properties of both; as fuch may be conceiv'd to change their Natures and States, without a total Diffolution of their Senfes by Death : And farther, it may poffibly be judged unpardonable in me not to point out every bleffed Abode, fuited to the Virtues, and all the various States an immortal Soul may be tranflated to ; but this is a Tafk above the human Capacity, or is the pure Province of Religion alone ; the Bufinefs of a Revelation rather than Reafon to difcover. Befides, it is enough for the prefent Purpofe, to prove, that Miriads of celeftial Manfions, are to be difcovered within our finite View, and by a kind of ocular Revelation, which vifibly extends the human Profpect, as it were, far beyond the Grave. It matters not whether a Race of Heroes fill thefe Worlds, or a Tribe of happy Lovers people thofe ; whether a Peafant in the Realms of *Orion* fhall ever become a Prince in the Regions of *Arcturus*, or a Patriarch in *Procion*, a Prophet in the *Precepæ*. Not to mention all the Stages human Nature may, or have been deftined to in any one World, as believ'd by the ancient Philofophers, befides the final Coalition of all Beings much more naturally to be expected in the *Sedes Beatorum*.

I fay, whatever our Cafe may be with regard to thefe *Queries* and Futurity, the Plan and Principles of this Theory will not be at all changed by it, fince what it is chiefly founded upon may be clearly demonftrated, fo clearly and inconteftably, that, with the Reverend Dr. *Young*, we may juftly conclude,

Devotion ! Daughter of Aftronomy !

and affirm with him alfo, That,

An indevout Aftronomer is mad.

But I find what I at firft propofed will prove too long for this Letter. However, I will endeavour to reward your Patience in my next, and continue, *&c.*

LETTER

III

LETTER the SIXTH.

Of General Motion amongſt the Stars, the Plurality of Syſtems, and Innumerability of Worlds.

SIR,

SINCE my laſt, you'll find by this, ſpeaking in the Stile of *Kercher*, that I have been very far from home, round almoſt the viſible Creation. I have indeed applied myſelf very cloſely to tranſcribe my Thoughts to you upon the old Subjeƈt the *Milky Way*, which my former Letter left imperfeƈted. To return then to the Theory of the Stars, and that yet unreconciled Phænomenon ; let us reaſon a little upon the viſible Order of the Stars in general, and ſee what Concluſions can be drawn from what every Aſtronomer knows of them, and cannot be diſputed.

First then, that the Stars are not infinitely diſperſed and diſtributed in a promiſcuous Manner throughout all the mundane Space, without Order or Deſign, is evident beyond a Doubt from this vaſt colleƈtive Body of Light, ſince no ſuch Phænomenon could poſſibly be produced by Chance, or exhibited without a deſigned Diſpoſition of its conſtituent Bodies.

If any regular Order of the Stars then can be demonſtrated that will naturally prove this Phænomenon to be no other than a certain Effeƈt ariſing from the Obſerver's Situation, I think you muſt of courſe grant ſuch a Solution at leaſt rational, if not the Truth ; and this is what I propoſe by my new Theory.

To a Speƈtator placed in an indefinite Space, all very remote Objeƈts appear to be equally diſtant from the Eye ; and if we judge of the *Via Laƈtea* from Phænomena only, we muſt of courſe conclude it a vaſt Ring of Stars, ſcattered promiſcuouſly round the celeſtial Regions in the Direction of a perfeƈt Circle.

But when we conſider the explanick Poſition of many other Stars, all of the ſame Nature, and not leſs numerous, together forming the great Sphere of Heaven, we generally find ourſelves quite at a Loſs how to reconcileth e two apparent Claſſes ; and I know none who have ever been ſucceſsful enough to reduce them to any one general Order.

You'll

You'll fay probably how fhall we make this chaofic Difpofition of the primary Luminaries agree with the fecondary Laws, and the juft Harmony obferved in the third * Creation, &c.

The Work now you fee is undertaken, and chiefly at your own Requeft, therefore I have a Right to expect you'll be very indulgent to the Author, and pafs over all his Faults, and allow him free Argument in Purfuit of thefe important Truths, which will in the End open perhaps a much wider Field of Contemplation to us, than at firft could be fuppofed to be intended by the *Genefis* of *Mofes*.

That Defcription of the Beginning of Nature is not without its Beauty and Noblenefs, fuitable to the Dignity both of the Author and Subject. But fhould we even in this knowing Age of the World pretend to account for the Original of Things, as *Mofes* to fupport his believed divine Legation, was obliged in fome meafure to do, we fhould foon be reduced to talk in the fame Stile, and perhaps with lefs Probability, than then at leaft appeared in his elegant Account of the Origin of the Univerfe, efpecially if we do but confider, that what he wrote, was only to the Senfes of a People who had not yet learnt to make ufe of their Reafon any other way, but from the Appearance of Things, and upon a Subject too fublime for vulgar Capacities in any Age, and had only been attempted in the deepeft Learning of *Egypt*, which, he though well acquainted with, the Generality of them were totally Strangers to.

In the firft Place it muft be granted, that the Stars being all of the fame Nature, are either all immoveable, or all fixed, that is all governed by one and the fame Principle.

Now to fuppofe them all fixed, and difperfed in an endlefs Diforder thro' the infinite Expanfe, which has long been the Opinion of many very able Aftronomers amongft the Antients, and even now received by too many of the Moderns, implies an Inactivity in thofe vaft and principal Bodies, fo much the Reverfe of what may be expected, and what we daily obferve through all the reft of their Attendants, namely, their own refpective Satellites, that we cannot poffibly upon any rational Grounds, advance one fingle Argument to fupport fo much as a Conjecture towards it, without betraying the greateft Simplicity, and next to an Affirmation reduce the whole Frame of Nature, and all corporeal Beings to a wild unmeaning Chance, arifing from an unnatural Difcord and Confufion.

For upon the Principles of Locality and Materiality, you having allowed me the Ufe of my Senfes and Reafon, as abfolutely neceffary towards conceiving any Idea of our prefent State, or of Futurity: Upon

---

* The Moon, Satellites of *Saturn* and *Jupiter*, &c.

thefe

thefe Principles I fay, unlefs our Faculties are ufelefs, if there are no other Bodies or Beings in the Univerfe than what we fee, and are now fenfible of, we muft now at the Height of this our prefent State, be as near Perfection as we can reafonably expect, and as fuch ourfelves the fupreme Beings of all Beings. To what End then do we form Ideas of a fucceeding Life, where a more exalted State cannot be hoped for.

How abfurd and impious this is I leave to your own Reafon and Reflection: This is the fatal Rock upon which all weak Heads and narrow Minds are loft and fplit upon, confequently ought to be the moft carefully avoided, not only as the Nurfe of Atheifm, but as the dreadful Father of Defpair: " For, fay they, thefe unhappy Wretches, to be always the " fame, is inconfiftent with a Change; and to be lefs than what we are, " any where hereafter, is full as difficult to conceive as to be more." Thus, unlefs we admit of fuperior Seats and much more glorious Habitations than thefe we are fenfible of, we ftrike at the very Root of a fair flourifhing Tree of Immortality, and muft become Authors of our own Defpair. I have often wonder'd how thinking Men could poffibly fall into fo grofs an Error, as that of a Spirit's Annihilation; and I fhould be glad to afk one of thofe fruitlefs Students, whether, upon the Evidence of our prefent Being, it is not much more rational, to hope for a future, than to expect a *Ne plus ultra* upon no Evidence at all. The Affirmative is certainly much more natural to be conceiv'd than the Negative. But if Chance were the Cafe, and that Chance produced all thefe regular and wondrous Works, 'tis to be wifhed at leaft, that Chance might do the fame again; and if not Chance, of courfe an eternal Direction: But Chance only can effect Diforder, Difcord, and Confufion; *ergo*, the vifible Harmony and Beauty of the Creation declare for a Direction; and this muft of Confequence, from its perfect Nature, proceed from the Wifdom and Power of an eternal Being, *God of Infinity*, the Author of all Ideas: And if this primitive Power produced us his Creatures from nothing, nothing can be wanting to revive our Frames again; and if from fomething, that fomething muft remain to eftablifh us in a future Life. But to return, how abfurd it is to fuppofe one Part of the Creation regular, and the other irregular, or a vifible circulating Order of Things, to be mixed with Diforder, and circumfcribing Part of an endlefs Confufion, is obvious to the weakeft Underftanding, and confequently we may reafonably expect, that the *Via Lactea*, which is a manifeft Circle amongft the Stars, confpicuous to every Eye, will prove at laft the Whole to be together a vaft and glorious regular Production of Beings, out of the wondrous Will or Fecundity of the eternal and infinite *one* felf-fufficient Caufe; and that all its Irregularities are only fuch as naturally arife from our excentric View: To demonftrate

which

PLATE XVII.

which abfolutely and inconteftibly, we fhall only want this one *Poftulata* to be granted, *viz. That all the Stars are, or may be in Motion:* This, if one may be allowed to judge of the Whole by the Similitude and Government of its Parts, I am perfwaded you will think a very reafonable Affumption; but that you may imbibe a good Opinion of this Affumption, and entirely come into this much better to be wifhed Hypothefis, I would have you confult thefe following Arguments.

First, it is allowed, as I have endeavoured to fhew, by all modern Philofophers, that the Sun and Stars are all of the fame or like Nature; confequently, that the Stars are all Suns, and that the Sun himfelf is a Star.

PLATE XVII.

Reprefents a kind of perfpective View of the vifible Creation, wherein A reprefents the Syftem of our Sun, B, that fuppofed round *Syrius,* and C, the Region about *Rigel.* The reft is a promifcuous Difpofition of all the Variety of other Syftems within our finite Vifion, as they are fuppofed to be pofited behind one another, in the infinite Space, and round every vifible Star. That round every Star then we may juftly conjecture a fimilar Syftem of Bodies, governed by the fame Laws and Principles with this our folar one, though to us at the Earth for very good Reafons invifible *. Secondly,

The Sun is alfo obferved to have a Motion round his own Axis in about twenty-five Days. Now, fince all the other † Planets which move in Orbits round him, and are within our Obfervation, are found to have a like Rotation round their Axis, may we not as reafonably imagine, that that Power which was able to give the Sun a Motion round his Axis, could and would at the fame time, with adequate Eafe, give him alfo an orbitular one? and why not, fince no progreffive Mutability can either take from, or difturb the boundlefs Property of an Infinity; and befides, feeing to imagine him at reft, is to impofe fuch an unnatural Stagnation upon the eternal Faculty, quite repugnant to that imparable Power which we fuppofe ftands in need of neither Sleep nor Reft?

'Tis true, the Sun may be faid to be the Governor of all thofe Bodies round him; but how? no otherwife than he himfelf may be governed by a fuperior Agent, or a ftill more active Force; and methinks it is not a

---

* *Anaximines* believed the Stars to be of a fiery Nature; and that there were certain terreftrial Bodies that are not feen by us, carried together round them. *Stob. Ecl. Phyf.* cap. 25. *Pythagoras* affirmed, that every Star is a World, containing Earth, Air, and Æther.

† *Saturn, Jupiter, Mars, Venus,* the Earth, Moon, and *Mercury.*

H 2 little

little abfurd to fuppofe he is not, fince we have difcovered by undoubted Obfervations, that the fame gravitating Power is common to all; and that the Stars themfelves are fubject to no other Direction than that which moves the whole Machine of Nature.

Thirdly, From many Obfervations of the polar Points, and the Obliquity of the Earth's Equator to the Plane of her folar Orbit compared together, the Sun is very juftly fufpected to have changed his fidereal Situation; and this muft either arife from a Change in the Pofition of the Earth's diurnal Axis, or from a Removal of the Sun himfelf, out of the primitive Plane of the *Orbis Magnus*. I believe you are fo much of a Mathematician, as to know that if either of thefe Facts be allowed, the Confequence I want will follow. I fhall not therefore here enter into any farther Difpute about it; but I think it will be neceffary to fubmit fome Obfervations to your Confideration, that may convince you that there is a Motion fomewhere to be thus difcovered, and whether in the Sun, or in the Stars, or in both, I leave to your own Determination, but to affift your Imagination, I refer you to

PLATE XVIII.

The Globe S is here fuppofed to reprefent the Sun, having changed its Situation by a local Motion from A to C, and B reprefents the Globe of the Earth in a permanent Pofition, with its principal Points and Circles, refpecting the primitive Plane A, B, K. Now in Confequence of the Angle of Variation, A, B, C, it evidently appears that a new ecliptic Plane, will be produced, as C, B, and alfo a Variation in the greateft Declination of the Sun, North and South from the Line of the *Equator* D, L. Hence, as in this Figure, the Obliquity of the Poles P, N, and G, F, will naturally decreafe, and is fhewn in Quantity by the Line of Aberration H, I.

Here follows a Table of the Change obferved in the Obliquity of the Ecliptic by Aftronomers of different Ages.

A Table of the Obliquity of the Ecliptic.

Ante-Chrifti		9	1
124	ARATO - - - - - - - - - - - -	24	00
——	HIPARCHUS - - - - - - - - -	23	51 $\frac{1}{3}$
127	ERATOSTHENES - - - - - - -	23	51 $\frac{1}{2}$

Anno

118

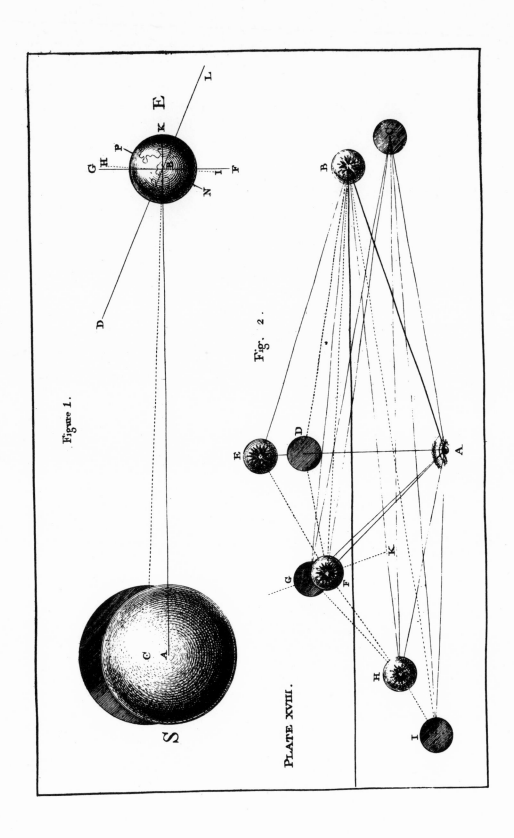

Figure 1.

Fig. 2.

PLATE XVIII.

Anno Dom.		o	′
140	PTOLOMY - - - - - - - - - -	23	51 ⅓
749	ABATEGNIUS - - - - - - - -	23	35 ½
1070	AIRAHEL - - - - - - - - -	23	34
1140	ALOMEAN - - - - - - - - -	23	33
1300	PROFATIOGRAD - - - - - - -	23	32
1458	PURBACCHIO - - - - - - - -	23	29 ½
1490	REGIOMONTAUS - - - - - - -	23	30
1500	COPERNICUS - - - - - - - -	23	28 ½
1592	TYCHO BRAHE - - - - - -	23	21 ½
1656	CASSINI - - - - -	23	29 ½

Now sure, if we consider this continual Decrease of the Sun's Declination, which can proceed from no other Cause than that of his having moved out of the primitive Plane; we need make no great Difficulty thus far, to think our Conjectures not irrational.

The following is a Citation from Dr. *Edmund Hally*, Astronomer-Royal. See *Philosophical Transactions*, Nᵒ. 355. p. 736.

" But while I was upon this Enquiry (*of the Obliquity of the Ecliptic*) I was surprized to find the Latitudes of three of the principal Stars in the Heavens, directly to contradict the supposed greater Obliquity of the Ecliptic, which seems confirmed by the Latitudes of most of the rest; they being set down in the old Catalogues, as if the Plane of the Earth's Orbit had changed its Situation amongst the fixed Stars, about 20′ since the Time of *Hipparchus*, particularly all the Stars in *Gemini* are put down, those to the Northward of the Ecliptic, with so much less Latitude than we find, and those to the Southward, with so much more southerly Latitude; and yet the three Stars *Palilicium*, *Sirius*, and *Arcturus*, do contradict this Rule: For by it, *Palilicium*, being in the Days of *Hipparchus*, in about 10 gr. of *Taurus*, ought to be about 15′ more southerly than at present, and *Sirius* being then in about 15 gr. of *Gemini*, ought to be 20′ more southerly than now; yet *Ptolomy* places the first 20′, and the other 22′ more northerly in Latitude than we now find them: Nor are these the Errors of Transcribers, but are proved to be right by the Declination of them set down by *Ptolomy*, as observed by *Timocharis*, *Hipparchus*, and himself; which shew, that these Latitudes are the same as those Authors intended. As to *Arcturus*, he is too near the Equinoctial Colure, to argue from him concerning the Change of the Obliquity of the Ecliptic; but *Ptolomy* gives him 33′ more North Latitude than .

than he is now found to have; and that greater Latitude is likewise confirmed by the Declinations delivered by the abovesaid Observations: So then these three Stars are found to be above half a Degree more southerly at this Time than the Antients reckoned them. When, on the contrary, at the same time, the bright Shoulder of *Orion*, has, in *Ptolomy* almost a Degree more southerly Latitude than at present, what shall we say then? It is scarce to be believed, that the Antients could be deceived in so plain a Matter, three Observers confirming each other. Again, these Stars being the most conspicuous in Heaven, are in all Probability the nearest to the Earth; and if they have any particular Motion of their own, it is most likely to be perceived in them, which in so long a Time as eighteen hundred Years, may shew itself by the Alteration of their Places, though it be intirely imperceptible in the Space of one single Century of Years: Yet, as to *Syrius*, it may be observed, that *Tycho Brahe* makes him 2 Min. more northerly than we now find him; whereas he ought to be above as much more southerly from his Ecliptic (whose Obliquity he makes $2'\frac{1}{2}$ greater than we esteem it at at present) differing in the Whole $4'\frac{1}{2}$.

One Half of this Difference may perhaps be excused, if Refraction were not allowed in this Case by *Tycho*; yet 2 Min. in such a Star as *Syrius*, is somewhat too much for him to be mistaken in.

But a more evident Proof of this Change is drawn from the Observation of the Application of the Moon to *Palilicium*, *An. Chris.* 509. *Mar.* 11. when, in the Beginning of the Night, the Moon was seen to follow that Star very near, and seemed to have eclipsed it, ἐπέβαλλε γὰρ ὁ ἀστηρ τῳ πᾶρα την διχοτομίαν μέρει τῆς κυ'ρτυς περιφειας τοῦ πεφωτισμένου μερους, *i. e. Stella apposita erat parti per quam bisecabatur limbus Lunæ illuminatus*, as *Bullialdus*, to whom we are beholden for this ancient Observation, has translated it. Now, from the undoubted Principles of Astronomy, this could never be true at *Athens*, or near it, unless the Latitude of *Palilicium* were much less than we at this Time find it *.

The Motion of *Arcturus* seems further confirmed, from the Observations of *Tycho Hevelius* and *Flamstead*; for *Hevelius* sets down the Distance of that Star from *Lyra* 4' greater than *Tycho* had observed it seventy-two Years before him, and *Flamstead* twenty-two Years after measured

* Vide *Bulialdi Astr. Philolaica*, p. 172.

† These are the nearest and greatest of the fixed Stars, the Motion of the others not having been observed, or being at too great a Distance, are either imperceptible, or have not been taken notice of.

the

the Diſtance betwixt the ſame two Stars, ſtill 3′ greater than *Hevelius* found it ; ſo that if *Lyra* had ſtood ſtill all that while, there was an Appearance of *Arcturus*'s having gone 7′ out of his Place in the Space of an hundred Years. See Dr. *Long*'s Aſtronomy, p. 274.

It is further to be obſerved, in Confirmation of the Motion of one of theſe Stars, that *Flamſtead* found the Diſtance of *Arcturus*, from the Head of *Hercules* 3′ greater than it is ſet down by the Prince of *Heſſe* ; and that his Diſtance from the *Lion's Tail* was a little decreaſed with 5′ ½ leſs Latitude than *Tycho* had obſerved. Hence, to make theſe Obſervations agree, one or both of them muſt have moved together equal to 7′. This Change of Place, which is quite contrary to all known Cauſes proceeding from the Earth, muſt therefore be occaſioned either by the Motion of the Sun, or by a particular Motion of their own ; but if, amongſt themſelves, they muſt all move, and if all be in Motion, the Sun muſt alſo move.

If theſe Obſervations, delivered down to us by very able Aſtronomers, be either true or near it, as great Allowances have been made for the Ignorance of the Ages in which they were taken, and the Inaccuracy of the Inſtruments, we may naturally conclude, that theſe Stars muſt have a Motion ; and if they move, as has been before obſerved, the Sun muſt alſo ; hence he cannot now be in the original Plane of the Earth's annual Direction, or at leaſt in the ſame identical Place he was at firſt poſſeſſed of : And if ſo, the Stars muſt alſo have the like Motion, though in different Directions, and all may thus be governed by the ſame impulſive Power.

To illuſtrate this primitive Motion of the Stars, and at the ſame time to ſhow that the Variety which appears in the Quantity of Motion can be no Objection to it,

See PLATE XVIII. *Fig.* 2.

Where A repreſents the Eye of an Obſerver, and B, E, F, H, various Syſtems, moving in different Directions thro' the mundane Space; it is evident that the Sphere B, having moved from C, and that of E, not having appeared to move at all, there muſt be a ſenſible Change in the new Poſition of theſe two Syſtems to one another, and ſo of the reſt; and tho' the apparent Motion of H, be much more than that of F, from the Point A, yet from C, they will appear leſs different, and from B, they will appear nearly equal. And farther, as the Direction from H, is in the Line I, H, and that of F, in the Line K, G, thoſe two Syſtems will appear to approximate, and the Magnitude of the Star in the firſt will be increaſed, creafed, and in the latter diminifhed. Thus, many of the Stars in the oldeft Catalogues, which were faid to be of the fecond Magnitude, are now become of the firft, and feveral of the firft are now judged to be of the fecond, &c.

But as this apparent Motion of the Stars at the Earth, muft, from its Nature, be very fmall, fo as fcarce to be difcovered in fome of them in lefs than an Age, with any Inftrument by the niceft Obferver, I judge it will be extremely proper in this Place to propofe fome Method, by which, in procefs of Time, the Truth of the Theory may be afcertained. The Way I think moft likely to fucceed is this.

PLATE XIX.

Is a Plan of the principal Stars that form the PLEIADES, correctly taken by a Combination of Triangles, as in the Figure, from whence it will naturally follow, all the whole Form being comprehended in much lefs than one Degree. That the moft minute local Motion in any one of thofe Stars in a very few Years, will be made fenfible to an Eye at the Earth. For Inftance, if any of the Stars that form the Letter A, or T, within the Term of ten or twenty Years, be found in the leaft to deviate from the Lines of their prefent Pofition and Direction, it will be evident beyond a Contradiction, that they have a Motion amongft themfelves, and fince at fuch a Diftance they cannot poffibly be affected by the Earth, it muft be a Motion of their own; and thus if any one can be proved, to change its Situation, with regard to the reft, we can have no new Difficulty in concluding that they all may do the fame.

Thus if any of the regular Triangles M B Z, Z P H, A Z M, T A Γ, or Π O I, &c. in due Time be carefully noted, we may venture to fay with great Safety, that the thoufandth Part of a Degree will be plainly difcovered.

PLATE XX.

Is a true Plan and Combination of the principal Stars that form the PERSEDES, in which other Obfervations may be made in a different Part of the Heavens, and perhaps with an Opportunity of being ftill more exact, the Areas of thefe Triangles, particularly that of Θ I K, and thofe of ρ and δ, being much lefs than the former, where the leaft Alteration poffible muft render them fenfibly diftorted. But here it muft be confidered, that the real Motion of the Stars, as well as their apparent, may be, and in all

PLATE. XIX.

PLATE. XX.

all Likelihood, is extreamly flow, for the moft minute, vifible, local Motion, will anfwer all the Purpofes we know in Nature, and the greateft feems to be that of the projectile, or centrifugal Force, which not only preferves them in their Orbits, but prevents them from rufhing all together, by the common univerfal Law of Gravity, which otherwife, as a finite Diftribution of either regular or irregular Bodies, they muft at length do by Neceffity.

I muft now inform you, that the above Obfervations were compleated in the Autumn Season, 1747, and were taken by myfelf; the Letters A, T, in *Plate* XIX, and the W in the XXth, as you may fee, having a very near Refemblance, or Similitude, to the Order thefe Stars are found to be in, together with the *Greek* Alphabet, I judged neceffary, by way of *Afterifm* and *Nomenclatura*, in cafe fuch fhould be wanted, as *Data* in future Difcoveries.

I come now to the principal Point in Queftion, which is to find a regular Difpofition of the Stars amongft themfelves, which will naturally folve both their general and particular Phænomena, efpecially the *Nebula* and *Milky Way*.

I am now, &c.

I   **LETTER**

LETTER THE SEVENTH

The Hypothesis, or Theory, fully explained and demonstrated, proving the sidereal Creation to be finite.

SIR,

I KNOW you are an Enemy to all Sorts of Schemes where they are not absolutely necessary, and may possibly be avoided; and for that Reason I have purposely omitted many geometrical Figures, and other Representations in this Work, which might have been inserted and in some Places, especially here I might have introduced Diagrams, perhaps more explicit than Words; but as you have frequently observed, they are only of Use to the few Learned, and contribute more to the taking away the little Ideas and Knowledge the more ignorant Many may be endued with, by a prejudicial Impression of imperfect Images, rather than the adding any new Light to their Understanding, I have purposely avoided, as much as possible, both here and every where, all such complex Diagrams as might be in Danger of betraying any the least such conscious Diffidence in you, arising from the Want of a proper *Precognita* in the Sciences.

This Imperfection, much to be lamented, as greatly to the Disadvantage of all mathematical Reasoning, I would willingly always prevent, in my Readers, and to chuse in my Friend; I shall therefore content myself with referring you to a few orbicular Figures, concave and convex, as may best suggest to your Fancy the simplest Way, a just Idea of the Hypothesis I have fram'd, and naturally enough I hope, render my Theory so intelligible, as to help you sufficiently to conceive the Solution aimed at, of the important Problem I have attempted.

As I have said before, we cannot long observe the beauteous Parts of the visible Creation, not only those of this World on which we live, but also the Myriads of bright Bodies round us, with any Attention, without being convinced, that a Power supreme, and of a Nature unknown to us, presides in, and governs it.

The

The Courfe and Frame of this vaft Bulk, difplay
A Reafon and fix'd Law, which all obey.

SHER. MANILIUS.

And notwithftanding the many wonderful Productions of Nature in this
our known Habitation, yet the Earth, when compared with other Bodies of our own Syftem, feems far from being the moft confiderable in it;
and it appears not only very poffible, but highly probable, from what has
been faid, and from what we can farther demonftrate, that there is as
great a Multiplicity of Worlds, varioufly difperfed in different Parts of the
Univerfe, as there are variegated Objects in this we live upon. Now, as
we have no Reafon to fuppofe, that the Nature of our Sun is different
from that of the reft of the Stars; and fince we can no way prove him fuperior even to the leaft of thofe furprifing Bodies, how can we, with
any Shew of Reafon, imagine him to be the general Center of the whole,
i. e. of the vifible Creation, and feated in the Center of the mundane
Space? This, in my humble Opinion, is too weak even for Conjecture,
their apparent Diftribution, and * irregular Order argue fo much a-
gainft it.

The Earth indeed has long poffeffed the chief Seat of our Syftem, and
peaceably reigned there, as in the Center of the Univerfe for many Ages
paft; but it was human Ignorance, and not divine Wifdom, that placed
it there; fome few indeed from the Beginning have difputed its Right to
it, as judging it no way worthy of fuch high Eminence. Time at length
has difcovered the Truth to every body, and now it is juftly difplaced
by the united Confent of all its Inhabitants, and inftead of being thought
the moft majeftick of all Nature's lower Works, now rather difgraces the
Creation, fo much it is reduced in its prefent State from what it had
Reafon to expect in the former.

Now it is no longer the only terreftrial Globe in the Univerfe, but is
proved to be one of the leaft Planets of the folar Syftem, and furprizingly
inferior to fome of its Fellow Worlds. The Sun, or rather the Syftem,
has almoft as long ufurped the Center of Infinity, with as little Pretence
to fuch Pre-heminence; but now, Thanks to the Sciences, the Scene be-
gins to open to us on all Sides, and Truths fcarce to have been dreamt of,
before Perfons of Obfervation had proved them poffible, invades our Senfes

---

* See the Zodaical Conftellations, you'll find that in fome Signs there are feveral Stars of
the firft, fecond, and third Magnitude, and in many others none of thefe at all.

I 2

with

with a Subject too deep for the human Underſtanding, and where our very Reaſon is loſt in infinite Wonders. How ought this to humble every Mind ſuſceptible of Reaſon!

In this Place, I believe, you will pardon a Digreſſion; which, in Anſwer to Part of your laſt Letter, I judge will not be very impertinent, tho' perhaps juſt here I cannot ſo well juſtify it.

Your late Converſation with our Friend Mr. * * *, I am perſwaded, muſt have been very entertaining; but I cannot help thinking his Reflections upon the Wonders of Nature and the Wiſdom of Providence, though I muſt allow them all to be very juſt and curious, inſtead of elevating the Mind to the Pitch he would have it, rather as conſidered above, depreſs it below the proper, nay I might ſay neceſſary, Standard of human Ideas.

This, probably, you'll ſay is an odd Turn, and may want ſome Explanation, ſince every Object in the Chain of Nature, muſt of Force be granted, a Subject worthy of our Speculations, being all together made, as in the Maximum of Wiſdom: But what I mean is this, ſince nothing is more natural for Beings in every State in ſearch after their own Advantages, and the Enlargement of their Ideas to look upward, ſure it may be preſumed, that Time may be miſpent, if not loſt in inſpecting too narrowly Things ſo little benefical in States below us; as Mr. *Pope* ſays,

> Why has not Man a microſcopic Eye?
> For this plain Reaſon, Man is not a Fly.
> Say what the Uſe, were finer Opticks given,
> To inſpect a Mite, not comprehend the Heav'n.

> *Eſſay on Man.*

Amuſement alone can never be ſuppoſed to be the ſole End of human Life, where even true Happineſs is a Thing we rather taſte than enjoy. The Mind we find capable of much more rational Pleaſure than can poſſibly fall within the Reach of human Power, either to promiſe or procure it; but then this very Defect in our preſent State of Exiſtence affords us no leſs than a moral Aſſurance, that ſome where in a future, we may, if we pleaſe, be entitled to the very *Plenum* of all Enjoyments.

The peculiar Buſineſs then of the human Mind naturally precedes its Amuſements, as evidently ordained to ſoar above all the inferior Beings of this World; and however our Natures may, thro' Indolence, or thro' Ignorance, degenerate, that of the Man can never be ſuppoſed to ſink into the Mole.

The propereſt Way then ſure for Men to preſerve their Pre-heminence over the Brute Creation, is to make uſe of that Reaſon and Reflection, which

which fo manifeftly diftinguifhes their natural Superiority. A right Application of which, muft of courfe then direct us to a forward, rather than a backward Search in the vaft vifible Chain of our Exiftence, which clearly connects all Beings and States as under the Direction of one fupreme Agent.

This is all I would have underftood by the foregoing Pofition, which, in one Word, implies no more than that the fublime Philofophy ought in all Reafon to be preferred to the Minute; but I hope you will not infer from this my feeming Partiality for the celeftial Sciences, that I mean to infinuate, that the Study of terreftrial Phyficks is not a rational Amufement.

Mr. ***, you fay, feems to lament the Tafte of Mankind in general much in the fame Degree as you do his I readily grant you; a Man who can talk fo well upon an Ant, might make a more entertaining Difcourfe upon the Eagle; but I beg his Pardon, and though we are all too ready, and moft apt to condemn all fuch Pleafures as vain or trifling, which we have no Share in, or Tafte for ourfelves; yet I don't think it follows, that thofe ingenious Labours of his are ufelefs. The Pleafures arifing from natural Philofophy are all undoubtedly great ones, whether we confider Nature in her higheft, or in her loweft Capacity; the Beauties of the Creation are every Day varied to us below, as much they are every Night above, and in both Cafes, through every Object, the Creator fhines fo manifeft, that we may juftly confider him every where fmiling full in the Face of all his Creatures, commanding as it were an awful Reverence, and Refpect, due not only to his Omnipotency, but alfo to his infinite Goodnefs and endlefs Indulgencies. This is the only Return our Gratitude can make for all thofe Bleffings he daily beftows upon us, and to this great Author of her Laws, Nature herfelf cries aloud through Myriads of various Objects, and after her own expreffive and peculiar Manner, feems to command us with an attractive Grace, to obferve her Sovereign, and admire his Wifdom. The Majefty, Power, and Dominion of GOD is beft difplayed in the external Direction of Things, his Wifdom and vifible Agency in the internal: Hence, by proper Objects, felected from both, attended with juft Reflections, we may certainly raife our Ideas almoft to the Pitch of Immortals; but how far the human Imagination may poffibly go, or how much Minds like ours may be improved, is a Queftion not eafily determined; but as natural Knowledge evidently increafes daily, and aftronomical Enquiries are the moft capable of opening our Minds, and enlarging our Conception, of confequence they muft be moft worthy our Attention of all other Studies. But of this I have faid enough, and think it is now more than Time to attempt the remaining Part of my Theory.

When

133

When we reflect upon the various Aspects, and perpetual Changes of the Planets, both with regard to their * heliocentric and geocentric Motion, we may readily imagine, that nothing but a like eccentric Position of the Stars could any way produce such an apparently promiscuous Difference in such otherwise regular Bodies. And that in like manner, as the Planets would, if viewed from the Sun, there may be one Place in the Universe to which their Order and primary Motions must appear most regular and most beautiful. Such a Point, I may presume, is not unnatural to be supposed, altho' hitherto we have not been able to produce any absolute Proof of it. See *Plate* XXV.

This is the great Order of Nature, which I shall now endeavour to prove, and thereby solve the Phænomena of the *Via Lactea*; and in order thereto, I want nothing to be granted but what may easily be allowed, namely, that the *Milky Way* is formed of an infinite Number of small Stars.

Let us imagine a vast infinite Gulph, or Medium, every Way extended like a Plane, and inclosed between two Surfaces, nearly even on both Sides, but of such a Depth or Thickness as to occupy a Space equal to the double Radius, or Diameter of the visible Creation, that is to take in one of the smallest Stars each Way, from the middle Station, perpendicular to the Plane's Direction, and, as near as possible, according to our Idea of their true Distance.

But to bring this Image a little lower, and as near as possible level to every Capacity, I mean such as cannot conceive this kind of continued Zodiack, let us suppose the whole Frame of Nature in the Form of an artificial Horizon of a Globe, I don't mean to affirm that it really is so in Fact, but only state the Question thus, to help your Imagination to conceive more aptly what I would explain *. *Plate* XXIII. will then represent a just Section of it. Now in this Space let us imagine all the Stars scattered promiscuously, but at such an adjusted Distance from one another, as to fill up the whole Medium with a kind of regular Irregularity of Objects. And next let us consider what the Consequence would be to an Eye situated near the Center Point, or any where about the middle Plane, as at the Point A. Is it not, think you, very evident, that the Stars would there appear promiscuously dispersed on each Side, and more and more inclining to Disorder, as the Observer would advance his Station towards either Surface, and nearer to B or C, but in the Direction of the general Plane towards H or D, by the continual Approximation of the visual Rays, crowding together as at H, betwixt the Limits D and G, they must in-

* Not to mention their several Conjunctions and Apulces to fixed Stars, *&c.* see the State of the Heavens in 1662, *December* the first, when all the known Planets were in one Sign of the Zodiac, *viz. Sagittarius.*

fallibly

PLATE.XXI.

PLATE XXII.

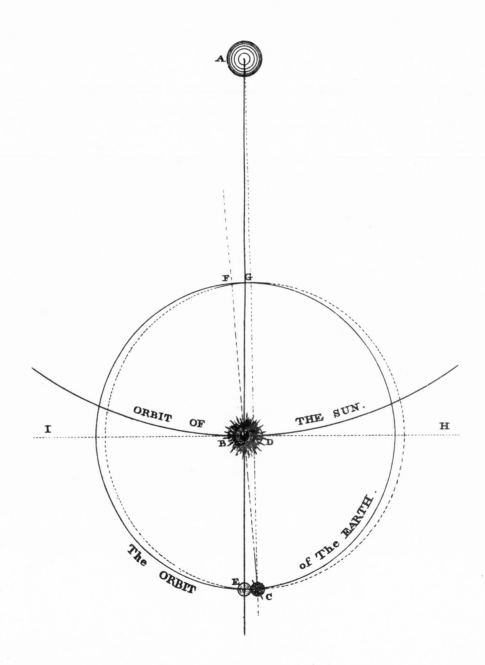

fallibly terminate in the utmoſt Confuſion.   If your Opticks fails you before you arrive at theſe external Regions, only imagine how infinitely
greater the Number of Stars would be in thoſe remote Parts, ariſing thus
from their continual crowding behind one another, as all other Objects
do towards the Horizon Point of their Perſpective, which ends but with
Infinity : Thus, all their Rays at laſt ſo near uniting, muſt meeting in the
Eye appear, as almoſt, in Contact, and form a perfect Zone of Light ; this
I take to be the real Caſe, and the true Nature of our *Milky Way*, and
all the Irregularity we obſerve in it at the Earth, I judge to be intirely
owing to our Sun's Poſition in this great Firmament, and may eaſily be
ſolved by his Excentricity, and the Diverſity of Motion that may naturally
be conceived amongſt the Stars themſelves, which may here and there, in
different Parts of the Heavens, occaſion a cloudy Knot of Stars, as perhaps
at E.

But now to apply this Hypotheſis to our preſent Purpoſe, and reconcile
it to our Ideas of a circular Creation, and the known Laws of orbicular
Motion, ſo as to make the Beauty and Harmony of the Whole conſiſtent
with the viſible Order of its Parts, our Reaſon muſt now have recourſe to
the Analogy of Things. It being once agreed, that the Stars are in Motion,
which, as I have endeavoured in my laſt Letter to ſhew is not far from
an undeniable Truth, we muſt next conſider in what Manner they move.
Firſt then, to ſuppoſe them to move in right Lines, you know is contrary
to all the Laws and Principles we at preſent know of ; and ſince there are
but two Ways that they can poſſibly move in any natural Order, that is,
either in right Lines, or in Curves, this being one, it muſt of courſe be
the other, *i. e.* in an Orbit ; and conſequently, were we able to view them
from their middle Poſition, as from the Eye ſeated in the Center of *Plate*
XXV. we might expect to find them ſeparately moving in all manner of
Directions round a general Center, ſuch as is there repreſented. It only now
remains to ſhew how a Number of Stars, ſo diſpoſed in a circular Manner
round any given Center, may ſolve the Phænomena before us.   There are
but two Ways poſſible to be propoſed by which it can be done, and one
of which I think is highly probable ; but which of the two will meet
your Approbation, I ſhall not venture to determine, only here incloſed I
intend to ſend you both.   The firſt is in the Manner I have above de-
ſcribed, *i. e.* all moving the ſame Way, and not much deviating from the
ſame Plane, as the Planets in their heliocentric Motion do round the
ſolar Body.   In this Caſe the primary, ſecondary, and tertiary conſtituent
Orbits, *&c.* framing the Hypotheſes, are repreſented in *Plate* XXII, and the
Conſequence of ſuch a Theory ariſing from ſuch an univerſal Law of Motion.

tion in *Plate* XXIII. where B, D denotes the local Motion of the Sun in the true *Orbis Magnus*, and E, C that of the Earth in her proper secondary Orbit, which of course is supposed, as is shewn in the Figure to change its sidereal Positions, in the same Manner as the Moon does round the Earth, and consequently will occasion a kind of Procession, or annual Variation in the Place of the Sun, not unlike that of the Equinoxes, or Motion of all the Stars together, from West to East round the Ecliptic Poles, and probably may in some Degree be the Occasion of it. This Angle is represented, but much magnified, by the Lines F, C, G, and the Unnaturalness, or Absurdity of a right Line Motion of the Sun by the Line I, H.

The second Method of solving this Phænomena, is by a spherical Order of the Stars, all moving with different Direction round one common Center, as the Planets and Comets together do round the Sun, but in a kind of Shell, or concave Orb. The former is easily conceived, from what has been already said, and the latter is as easy to be understood, if you have any Idea of the Segment of a Globe, which the adjacent Figures, will, I hope, assist you to. The Doctrine of these Motions will perhaps be made very obvious to you, by inspecting the following Plates.

PLATE XXIV.

Is a Representation of the Convexity, if I may call it so, of the intire Creation, as a universal Coalition of all the Stars consphered round one general Center, and as all governed by one and the same Law.

PLATE XXV.

Is a central Section of the same, with the Eye of Providence seated in the Center, as in the virtual Agent of Creation.

PLATE XXVI.

Represents a Creation of a double Construction, where a superior Order of Bodies C, may be imagined to be circumscribed by the former one A, as possessing a more eminent Seat, and nearer the supream Presence, and consequently of a more perfect Nature. Lastly,

PLATE XXVII.

Represents such a Section, and Segments of the same, as I hope will give you a perfect Idea of what I mean by such a Theory.

Fig. 1. is a corresponding Section of the Part at A, in *Fig.* 2. whose versed Sine is equal to half the Thickness of the starry Vortice A C, or B A. Now I say, by supposing the Thickness of this Shell, 1. you may imagine the middle Semi-Chord A D, or A E, to be nearly 6; and con-

sequently,

PLATE XXI.

PLATE XXIV.

PLATE XXV.

PLATE XXVI.

PLATE XXVII.

Figure 1.

Fig. 2.

PLATE XXVIII.

Figure I.

Fig. II.

Fig. III.

PLATE. XXIX.

Figure I.

Fig. II.

thus in a like regular Diftribution of the Stars, there muft of courfe be at leaft three Times as many to be feen in this Direction of the Sine, or Semi-chord A E, itfelf, than in that of the femi-verfed Sine A C, or where near the Direction of the Radius of the Space G. *Q. E. D.*

But we are not confined by this Theory to this Form only, there may be various Syftems of Stars, as well as of Planets, and differing probably as much in their Order and Diftribution as the Zones of *Jupiter* do from the Rings of *Saturn*, it is not at all neceffary, that every collective Body of Stars fhould move in the fame Direction, or after the fame Model of Motion, but may as reafonably be fuppofed as much to vary, as we find our Planets and Comets do.

Hence we may imagine fome Creations of Stars may move in the Direction of perfect Spheres, all varioufly inclined, direct and retrograde; others again, as the primary Planets do, in a general Zone or Zodiack, or more properly in the Manner of *Saturn*'s Rings, nay, perhaps Ring within Ring, to a third or fourth Order, as fhewn in *Plate* XXVIII. nothing being more evident, than that if all the Stars we fee moved in one vaft Ring, like thofe of *Saturn*, round any central Body, or Point, the general Phænomena of our Stars would be folved by it; fee *Plate* XXIX. *Fig.* 1. and 2. the one reprefenting a full Plane of thefe Motions, the other a Profile of them, and a vifible Creation at B and C, the central Body A, being fuppofed as *incognitum*, without the finite View; not only the Phænomena of the *Milky Way* may be thus accounted for, but alfo all the cloudy Spots, and irregular Diftribution of them; and I cannot help being of Opinion, that could we view *Saturn* thro' a Telefcope capable of it, we fhould find his Rings no other than an infinite Number of leffer Planets, inferior to thofe we call his Satellites: What inclines me to believe it, is this, this Ring, or Collection of fmall Bodies, appears to be fometimes very excentric, that is, more diftant from *Saturn*'s Body on one Side than on the other, and as vifibly leaving a larger Space between the Body and the Ring; which would hardly be the Cafe, if the Ring, or Rings, were connected, or folid, fince we have good Reafon to fuppofe, it would be equally attracted on all Sides by the Body of *Saturn*, and by that means preferve every where an equal Diftance from him; but if they are really little Planets, it is clearly demonftrable from our own in like Cafes, that there may be frequently more of them on one Side, than on the other, and but very rarely, if ever, an equal Diftribution of them all round the *Saturnian* Globe.

How much a Confirmation of this is to be wifhed, your own Curiofity may make you judge, and here I leave it for the Opticians to determine. I fhall content myfelf with obferving that Nature never leaves us without

K   a fuffia fufficient Guide to conduct us through all the neceffary Paths of Know-ledge ; and it is far from abfurd to fuppofe Providence may have every where throughout the whole Univerfe, interfperfed Modules of every Creation, as our Divines tell us, Man is the Image of God himfelf.

Thus, Sir, you have had my full Opinion, without the leaft Referve, concerning the vifible Creation, confidered as Part of the finite Univerfe ; how far I have fucceeded in my defigned Solution of the *Via Lactea*, upon which the Theory of the Whole is formed, is a Thing will hardly be known in the prefent Century, as in all Probability it may require fome Ages of Obfervation to difcover the Truth of it.

It remains that I fhould now give you fome Idea of Time and Space ; but this will afford Matter fufficient for another Letter.

I am, &c.

LETTER

LETTER the EIGHTH.

Of Time and Space, with regard to the known Objects of Immensity and Duration.

S I R,

THE Opportunity you gave me in your laft Vifit, of fhewing you my general Scheme of the Univerfe, I find, befides the Pleafure it then gave, is now attended with many ufeful Advantages.

I now not only hope to be better underftood for the future, but have reafon to expect what I now write will merit your Attention more, and have fome Title to your Approbation. The Ideas I have fram'd of Time and Space, will now more gradually fill your Imagination both with Wonder and Delight, before they can arife fo high as to be loft in an Eternity and the Infinity of Space. And I am fully perfwaded your farther Inquiries into thefe vaft Properties of the Deity, will here be anfwered intirely to your Satisfaction. You muft allow me now to be in fome meafure a Judge of what I think will pleafe you moft, from the Obfervations you have made upon my general Syftem, or otherwife you would have reafon to think me perhaps too prefuming : But I flatter myfelf the great Difficulty is now over ; and what remains to be faid, will alfo naturally follow from what has gone before, that this Letter, I guefs, will go near to furnifh you with all the Ideas you wifh to form upon the Subject. To what you have faid of my having left out my own Habitation in my Scheme of the Univerfe, having travell'd fo far into Infinity as both to lofe fight of, and forget the Earth, I think I may juftly anfwer as *Ariftotle* did when *Alexander*, looking over a Map of the World, enquir'd of him for the City of *Macedon* ; 'tis faid the Philofopher told the Prince, That the Place he fought for was much too fmall to be there taken Notice of, and was not without fufficient Reafon omitted.

The Syftem of the Sun compar'd but with a very minute Part of the vifible Creation, takes up fo fmall a Portion of the known Univerfe, that in a very finite View of the Immenfity of Space, I judg'd the Seat of the Earth to be of very little Confequence, could I have poffibly reprefented it, as not only being one of the fmalleft Objects in our Regions, but in a

K 2

manner

manner infinitely lefs than even her own annual Orbit, and had nothing to do with my main Defign, which was to reprefent all our planetary Worlds as one collective Body, and begin my comparative Scale of Magnitude from the Sun only and his Sphere of activity; as the fmalleft Object I could with any Propriety pretend to exprefs in fuch a Plan.

In fome Meafure to convince you that I have committed no Error in this, I will try by fome lefs mathematical Method than that of meer Numbers, to imprint an Idea in your Mind of the true Extent of the folar Syftem, and the Magnitude of all its moving Bodies, by natural Objects moft familiar to your Senfes. When we endeavour to form any Idea of Diftance, Magnitude, or Duration, by Numbers only, we fo foon exceed the Limits of Conception, that this way we find our Faculties of reafoning as finite as our Senfes; and no doubt 'tis right it fhould be fo, Providence, as it were, having ordain'd that the firft fhould only attend the laft, in fuch an adequate Degree to a determin'd Diftance; but what Diftance or Degree of Knowledge is deftin'd to human Nature, none but the Power that gave it can tell. 'Tis certain that beyond the third or fourth Place of our Nomenclator, we receive but very faint Impreffions of the thing expreft, and can frame fcarce any Notion at all of either Number, Diftance, or Magnitude, fignified beyond it: Hence Aftronomers are frequently oblig'd to have recourfe to mixt Ideas, and make Things of different Natures and Properties affift each other, to excite more adequate Ideas of what they would have conceived. Thus to exprefs immenfe Diftances and Magnitude, they frequently apply themfelves to Time and Motion; and *vice verfa*, to fignify a long Duration, they have often recourfe to Diftance and Matter, removing, in Imagination, Worlds of Sand, Grain after Grain, to fome remote known Region.

Hefiod, * to exprefs his Idea of the Diftance from his higheft Heaven to Earth, and from Earth to Hell, or *Tartarus*, fuppofes an Anvil to be let fall from one to the other, which he fays in nine natural Days would reach the Earth from Heaven, and in the fame time would fall from the Earth to Hell. † *Homer* makes his *Vulcan* fall from Heaven to the Ifland of *Lemnos* in much lefs Time, not exceeding one full artificial Day.

* From the high Heaven a brazen Anvil caft,
Nine Nights and Days in rapid Whirls would laft,
And reach the Earth the Tenth, whence ftrongly hurl'd;
The fame the Paffage to th' infernal World.

COOKE.

† Hurl'd headlong downward from th' etherial Height;
Tofs'd all the Day in rapid Circles round,
Nor till the Sun defeended touch'd the Ground.

POPE.

Modern

Modern Aftronomers have made ufe of the fwifteft Velocity of a Cannon-Ball as continued thro' the Space they would fo defcribe, and in this Light, the Diftance to the Sun has been by many compar'd to twenty-five Years Motion of a Cannon-Ball, fuppofing it to travel at the Rate of 100 Fathom in a Moment, *i. e. the Pulfe of an Artery*; and that a Journey fo performed to one of the neareft fix'd Stars, would take the fame Body at leaft 100,000 Years before it could arrive there. But the Method I have chofe to convey my Ideas of the Magnitude of the planetary Bodies, and the Extent of the vifible Creation to you, I am willing to hope you will find ftill more familiar, comprehenfive, and eafy: And it only depends upon your Remembrance of a very few known Objects, and their neighbouring Diftances, which may be prefumed you are, or have been, very well acquainted with. You have not only very lately but very often been in *London*, and muft, I think, retain fome Idea of the Dome of St. *Paul*'s, tho' I own I ought not to be forry if you fhould chance to have forgot it, provided it might prove a Means of making your Vifits more frequent. The Diameter of the Dome of this Church is 145 Feet: Now if you can imagine this to reprefent the Surface of the Sun, a fpherical Body 18 Inches diameter, will juftly reprefent the Earth in like Proportion; and another of only five Inches diameter, will reprefent the Moon. The Truths of thefe Proportions I have fhewn in my *Clavis Cœleftis*; and the Reafon why I have here fixt upon the Dome of this Church for my firft Object of Comparifon, will naturally appear from what follows.

From the Magnitude of the Earth on which we live, as from a known Scale with refpect to its Parts compar'd with our own Bodies, we naturally frame our firft Ideas of Extent, and fix our Rationale of Remotenefs; by which we are fufficiently enabled to judge of all other fenfible Diftances within one finite View. And hence by the undoubted Principles of Geometry, having firft given the Meafurement of the Earth in any known Proportion with any other Quantity moft familiar to our Senfes, and the Angle of Appearance, or Parallax to any perceivable Object, we can eafily find in homogenial Parts its true Diftance from the Eye. And thus allowing for fome fmall tho' unavoidable Errors, that may poffibly arife from the Difficulties of Obfervation (efpecially fmall Angles and minute Quantities) we can always determine to a fufficient, and very frequently to a juft Exactnefs, the relative Diftance of all vifible Bodies, remote or near, fuch as the Planets, Comets, and the Sun.

* In this Manner Aftronomers having procur'd a comparative Standard, reduc'd to fome known Meafure, as *Englifh* Miles, Leagues, Semi-Orbs or

Orbits,

* Parallax is the changeable Pofition of Bodies to different Situations of the Eye. Firft having found the Quantity of a Degree (*i. e.* a 6oth Part of the Circumference) upon the

Earth's

Orbits, with all the Force of analogical Reasoning, clearly can demonstrate the Place and Distance of any Object within the Reach of Observation, and judge of Distances almost indefinite.

PLATE XXX.

Will help you to very correct Ideas of the real Magnitude of the Globe of the Earth, compar'd with the just Extent of the Island of *Great-Britain*, which you will find with *Ireland*, and the rest of its Islands, seated near the Center of the Projection. This as a Standard will enable you to judge of all other Distances more perfectly; and first I shall consider that of the Sun.

The Sun is found to be mean distant from the Earth nearly 81 Millions of Miles, or 6877,5 Diameters of the Earth; and *Saturn*, the remotest Planet from him is at his greatest Distance from us about 858 Millions of Miles: Yet these Distances are but the beginning of Space, and only serve to open our Ideas for farther Search.

The great Comet of 1680, as I have some where said before, was found to move in so vast an excentrick Orbit, that in its aphelion Point it would be 14,4 Times as far from the Sun, as the Orbit of *Saturn*, and hence at least eleven thousand and two hundred Millions of Miles from us. Now since the wise Creator hath so dispos'd all the independent Parts of the Creation, such as the several Systems of primary and secondary Planets, &c. at so great a Distance from each other, that the Laws of any one in no wise shall interfere, disturb, or interrupt the Principles of another; this Comet, which we can easily prove belong'd to our own Sun, we may well imagine came not near any other; and tho' at that vast Distance from the solar Body, yet still there must have remain'd a Space sufficient to divide or seperate the sensible activity of neighbouring Systems, that they may not rush upon each other. Hence we may reasonably suppose, that the nearest Star can be no nearer than a triple Radius of its active Sphere; and provided they are all in regular Order, and much of the same Magnitude with one another (which no Arguments can possibly contradict) this Radius we may justly make 2000 times the Distance of our Earth. For admitting the utmost Limits of the Sun's Attraction to exceed this Sphere of the Comets, as far as the Sphere of the Comets

Earth's Surface, *Aratosthenes* discover'd that the Magnitude of the whole was easily known; and then from the Moon's horizontal Parallax having given the Radius of the Earth, the Distance of the Moon is soon determined; next by the menstrual Parallax of the Lunar Orbit, the Distance of the Sun is found; and by the Elongation of the inferior Planets, their mutual Distance from each other; and, lastly, from the annual Parallax of the Earth's Orbit, all the other Orbits of the superior Planets are easily found.

exceeds

THE
EARTH

PLATE XXX.

exceeds that of the Planets, which is nearly 14,4 times, the Radius of the folar Syftem will be extended every way 200 Radius's of the Orbit of *Saturn*, and confequently the Diftance from Star to Star will not be lefs than 6000 times the Radius of our *Orbis Magnus*, and confequently upwards of 480,000,000,000 Miles. That this is even lefs than the real Truth, and may be defended as a very moderate Computation, grounded upon Reafon, we have infallible Demonftration to witnefs, and make appear as thus.

We know from the Nature of Diftance and Motion that the Stars may have an annual Parallax, but it is fo very fmall, that the very beft Aftronomers have never yet been able to affign what the Quantity really is. Yet it is allow'd by univerfal Confent, that it can't poffibly be more that one Minute of a Degree, and may probably be much lefs. Mr. *Flamftead*, by repeated Obfervations, made it in fome of them upwards of 40″; but Mr. *Bradley* has endeavour'd to prove it is every where too fmall to be determined, and affigns this Angle to another Caufe. This way then we cannot make their Diftance lefs; and to prove that it is fomething more than I have faid it is, let us even increafe the doubtful Parallax of 40″ to the moft it poffibly can be, *viz.* to 60″ or 1′; and by the Solution of the Triangle, we fhall find that the neareft Star is 6875 times the Radius of the Earth's Orbit from the Sun: And this tho' more than any other Proportion makes them, is ftill undeniably lefs than the Truth, which every Mathematician will of courfe be convinc'd of; and you yourfelf of force muft believe, when you are told, that the fmaller the Angle of Parallax is, the farther the Body is remov'd from us. By which Rule, according to Mr. *Flamftead*'s Obfervations, the Diftance muft be ftill greater: By the optical Experiment of * Mr. *Huygins*, greater ftill than this; and according to Mr. *Bradley*, fo much more as not even too be determin'd.

Now if the reft are in general from each other, allowing the fame Extent of Syftem, and as much to part the like Extreams of active Virtue, be in fuch Proportion of aerial Space, it will appear, that to pafs from any one Star to another, we muft fly thro' fo vaft a Tract of pure Expanfe or Ether, that to vifit any one of the moft neighbouring Syftems, could we travel even as faft as the fwifteft Eagle flies, for Inftance, 500 Miles *per* Day, yet fhould we be 3,000,000 of Years upon our way before we could arrive there; and if continuing on to view the Regions of the reft within the known Creation, Myriads of Ages would be fpent, and yet we could not hope to fee the whole of but the fmalleft Conftellation.

* 27664 Radius's of the *Orbis Magnus*, equal to the Diftance of *Syrius*, whofe Parallax fhould be to anfwer it but 14″ 48‴.

But

But what Idea of Diftance can you receive from this fort of Eftimation, where Numbers arife fo very high. I own to you mine are foon quite loft by this Method of counting, either, Diftances or Duration. I believe few People can range their Ideas with fuch Perfpicuity, as to arrive at any adequate Notion of any Number above a thoufand.

To give you therefore a clearer Idea of Diftance, and imprefs the Proportions of Space more ftrongly and fully in your Mind, let us fuppofe the Body of the Sun, as I have faid before, to be reprefented by the Dome of St. *Paul's*; in fuch Proportion a fpherical Body eighteen Inches Diameter, moving at *Mary-le-bone*, will juftly reprefent the Earth, and another of five Inches Diameter, defcribing a Circle of forty-five Feet and a half Radius round it, will reprefent the Orbit and Globe of the Moon. A Body at the *Tower* of 9,7 Inches, will reprefent *Mercury*; and one of 17,9 Inches at St. *James's* Palace will reprefent the Planet *Venus*; *Mars* may be fuppofed at a Diftance, like that of *Kenfington* or *Greenwich*, 10 Inches Diameter; *Jupiter*, imagined to be at *Hampton-Court*, or *Dartford* in *Kent*; and *Saturn*, at *Cliefden*, or near *Chelmsford*: The firft reprefented by a Globe 15 Foot 4 Inches Diameter, the latter by one of 11 Feet ¼ and his Ring four Feet broad: Thefe would all naturally reprefent the planetary Bodies of our Syftem in their proper Orbits and proportional Magnitudes, as moving round the Cupola of St. *Paul's*, as their common Center the Sun. And preferving the fame natural Scale, the Aphelion of the firft Comet would be about *Bury*, the fecond at *Briftol*, and the third near the City of *Edinburgh*. But if you will take into your Idea one of the neareft Stars; inftead of the Dome of St. *Paul's*, you muft fuppofe the Sun to be reprefented by the gilt Ball upon the Top of it, and then will another fuch upon the Top of St. *Peter's* at *Rome* reprefent one of the neareft Stars.

The whole Syftem exhibited in the above Proportion, would be nearly as follows :

Diameter of the Sun   145 Feet.
      Saturn   11,587, his Ring 27,54, its Breadth 4.
      Jupiter, 15,39.
      Mars,   10,15 Inches.
   the Earth, 18,125.
      Venus, 17,98
      Mercury, 9,715
  and the Moon, 4,93

Diftance

* Diſtance of *Saturn* from the Sun, 27 Miles, and 1700 Yards.
    Jupiter,   15 Miles, and 458 Yards.
    Mars,   4 Miles, and 751 Yards.
    the Earth,   2 Miles, and 1632 Yards.
    Venus,   2 Miles, and 217 Yards.
    Mercury,   1 Mile, and 267 Yards.
and of the Moon, from us, 45 Yards and a half.

That of the moſt diſtant Comet 390, and the neareſt of the Stars not leſs than 6875, † Radius's of the *Orbis Magnus*.

Now, if like Creations crowd the vaſt Depths of Infinity, and if each are adapted to receive Beings of different Natures, where muſt our Wonder and Ideas have end?

As it is evident in the Sign *Taurus*, in *Perſeus*, and *Orion*, that we can plainly perceive Stars to the ſixth and ninth Magnitude, the former with our naked Eye, the other by the Help of Teleſcopes, the viſional ocular Creation cannot be leſs than 4,320,000,000,000 Miles in ſemi Diameter, and admitting a regular Diſtribution of thoſe primordial Bodies amongſt themſelves, the Depth, or moſt remote Limits of the *Vortex Magnus* from Side to Side, cannot be leſs than 8 m, m, 640 thouſand of Million of Miles, admitting it is no more than what we ſee; and laſtly, ſuppoſing our Syſtem to be ſituated nearly in the Middle of the *Vortex Magnus* (which, from the viſible Order of the Stars, we may juſtly conjecture, with the higheſt Probability of Truth) the neareſt Diſtance of the *Ens Primum*, in the Realms of eternal Day, will riſe to 30,000,000,000,000 Miles, but more probably to 100,000,000,000,000 Miles, making the Confines of Creation from Verge to Verge in the firſt Caſe, upwards of 68 Million of Millions of Miles, Diameter, and by the laſt above 200′. But, if we compute the Diſtance of the Stars after the Manner of *Huygens*, for his Diſtance of *Syrius* from the Sun, the Diſtance of the Region of Immortality without exceeding Probability may riſe to near 1,000,000,000,000,000 Miles.

Now to paſs by any progreſſive Motion from the outward Verge, or Borders of the Creation, thro' the ſtarry Regions of Mortality, if I may call

---

* Of the Satellites of *Saturn* in the above Proportion. And thoſe of *Jupiter*.

The 1 ⎫ ⎧ 27,96 ⎫   The 1 ⎫ ⎧ 28,51 ⎫
 2 ⎬ would be ⎨ 35,52 ⎬ Feet diſtant from his 2 ⎬ would be ⎨ 69,177 ⎬ Feet diſtant from him.
 3 ⎨ ⎬ 50, ⎨ Center.   3 ⎨ ⎬ 110,224 ⎨
 4 ⎬ ⎩ 114, ⎭   4 ⎭ ⎩ 190, ⎭
 5 ⎭ 341,9

             °   ′   ″
* Radius, or Sign of 89 59 30 —— —— 10,0000000
Sine ſubſtract of   0 0 30 —— —— 6,1626961
                     —————————
Hence the Diſtance 6875,5 — — 3,8373039

          L   them

them fo, as far as the Center of the *Ens Primum*, or *Sedes Beatorum*, accord-ing to *Homer*, or *Milton*'s Manner of meafuring Space, a Body falling, or a Being moving with a Velocity but of 1000 Feet *per* Minute, *i. e.* at the Rate of 20,000 Yards *per* Hour, or about 300 Miles *per* Day, would be at leaft 300,000,000 Years upon its Journey thither, if not 1,000, m, and perhaps much more, without offending Probability ; but even three Million Centuries, or Ages, fure is enough to be employ'd, in paffing from one Place to another ; therefore, we may conclude, the Soul muft have fome other Vehicle than can be found in the Ideas of Matter to convey it fo far, at leaft at once. Hence we may truly infer, that the Soul muft be immaterial, and that in all Probability there may be States in the Univerfe fo much more longer lived than ours, that, compared with the Age of Man, the Age of fuch Beings may be almoft as an Eternity, or rather, as that of the human Species to that of a Sun-born Infect.

Again, if there are ftill Stars beyond all thefe of other Denomination, which we do not here perceive, how vaftly muft thefe Numbers be in-creafed, to exprefs, almoft without Idea, the amazing Whole of this one vifible Creation ; but what has been already faid, I judge will be fufficient to fhow the Immenfity of Space, and help you to conceive the ftupendi-ous Nature of an endlefs Univerfe ; every where the home Poffeffion, Production, and inftantaneous Care, of an infinite good Being, perfectly wife, and powerful, of whom we can have no Idea more, than a Being in dark Privation can have of Light, but through the Luftre of his own re-fplendent Attributes.

Thus, having attempted to enlarge your Ideas of the Creation in general, and in fome meafure having confidered the Indefinity of Space, I fhall in the next Place proceed to give you fome Account of my Notions of Time.

As Diftance is the Meafure of Magnitude and of all Extent, and helps our Imagination to the Ideas of Space, fo are progreffive Moments the Meafure of Velocity, and makes us fenfible of Duration : And as Space may be ex-tended through all Infinity, fo Time may be continued as to Eternity. This Succeffion of temporal Ideas impreffed, or excited in the Mind, as an Effect of Matter in Motion, producing a perpetual Change, both of Ob-jects earthly and celeftial, enables us not only to reflect upon paft Viciffi-tudes of Nature, but from their regular Courfes, known Order and Returns, predict Phænomena to come, and prove the periodical Effects of Na-ture's conftant Laws fo juft and certain, that Time may be faid with Truth, to co-exift with Motion.

Meafure being a certain Quantity of Senfation interwove with our Ideas of Diftance and Duration, proceeding from a Reflection of what is im-preffed upon the Mind by fome external Object, I muft again return to our Mother of Ideas the Earth, and from thence, as I did, of Diftance,

frame

frame the original Images beſt ſuited to the Underſtanding, proper for our Judgment of Duration.

Time takes its firſt Denomination from the diurnal Rotation of the Earth upon its Axis, which we call a natural Day, and this for obvious Reaſons we ſubdivide in twenty-four Parts or Hours. This diurnal Motion having been ſucceſſively repeated, and the Day renewed three hundred and ſixty-five Times, we find that all the vegetable World has gone through all its Variegations, and Nature has again put on the ſame Face, adapted to the Seaſon ; during which Time, and indeed which occaſions this general Change and Repetition, the Earth is found to make one intire Revolution round the Sun. This Space, or Period of Time, we call a ſolar, or rather a natural Year ; and from our Senſibility of this, and its conſtituent Parts, both horary and diurnal, we form our general Judgment of Duration.

Saturn, the moſt remote, and moſt regular Planet in our Syſtem, as has been ſaid before, performs one Revolution round the Sun in about twenty-nine of the above ſolar Years: The great Comet of 1680 makes but one periodical Return in five hundred and ſeventy-five of thoſe Years, and the general Motion of the Stars, ariſing from the Proceſſion of the Equinoxes, altogether continually changing their Aſpect, or Poſition, at the Rate of 50″ *per* Year round the ecliptic Poles, compleats but one Revolution in 25920 Years ; in which Time the whole ſidereal Frame of Heaven has changed, and every Star returned to the ſame Point of the ſolar Sphere it ſet out from. This is by many called the great *Saturnian* Year : Concerning which, Mr. *Addiſon* has thus tranſlated an eminent Author.

> When round the great *Saturnian* Year has turn'd,
> In their old Ranks the wandering Stars ſhall ſtand,
> As when firſt marſhall'd by the Almighty's Hand.
>
> ADDISON.

Now, if this ſidereal Revolution, ariſing from a ſecondary Cauſe, require this Number of Years to perfect one Rotation, what muſt their primitive Orbits take to circumſcribe the *Vortex Magnus*.

It has been obſerved, that the biggeſt Star to us ſcarce moves a Minute in an hundred Years, and the moſt remote as inſenſibly for Ages, from whence and what has been already ſaid of the imagined Diſtance of the general Center, we may frame this probable and well-grounded Gueſs, that the mean Revolution of a Star near the Middle of the *Vortex Magnus*, cannot be made in leſs than a Million of Years, and though to us imperceptible, our Sun in his own orbicular Direction, may be moving many Miles *per* Day. Beſides, if local Motion can be proved amongſt the Stars, what leſs than an Eternity can again reſtore them to their original Order and primitive State.

L 2 Such

Such vaſt Room in Nature, as *Milton* finely expreſſes it, cannot be without its Uſe; and nothing but abſolute Demonſtration is wanting (which from their Nature and Diſtance cannot be expected) to confirm the grand Deſign, ſo ſuited to the Deity's infinite Capacity, and of eternal Benefit to all his Creatures, eſpecially Beings of a rational Senſe, and in particular Mankind.

Of theſe habitable Worlds, ſuch as the Earth, all which we may ſuppoſe to be alſo of a terreſtrial or terraqueous Nature, and filled with Beings of the human Species, ſubject to Mortality, it may not be amiſs in this Place to compute how many may be conceived within our finite View every clear Star-light Night. It has already been made appear, that there cannot poſſibly be leſs than 10,000,000 Suns, or Stars, within the Radius of the viſible Creation; and admitting them all to have each but an equal Number of primary Planets moving round them, it follows that there muſt be within the whole celeſtial Area 60,000,000 planetary Worlds like ours. And if to theſe we add thoſe of the ſecondary Claſs, ſuch as the Moon, which we may naturally ſuppoſe to attend particular primary ones, and every Syſtem more or leſs of them as well as here; ſuch Satellites may amount in the Whole perhaps to 100,000,000, or more, in all together then we may ſafely reckon 170,000,000, and yet be much within Compaſs, excluſive of the Comets which I judge to be by far the moſt numerous Part of the Creation.

In this great Celeſtial Creation, the Cataſtrophy of a World, ſuch as ours, or even the total Diſſolution of a Syſtem of Worlds, may poſſibly be no more to the great Author of Nature, than the moſt common Accident in Life with us, and in all Probability ſuch final and general Doom-Days may be as frequent there, as even Birth-Days, or Mortality with us upon the Earth.

This Idea has ſomething ſo chearful in it, that I own I can never look upon the Stars without wondering why the whole World does not become Aſtronomers; and that Men endowed with Senſe and Reaſon, ſhould neglect a Science they are naturally ſo much intereſted in, and ſo capable of inlarging the Underſtanding, as next to a Demonſtration, muſt convince them of their Immortality, and reconcile them to all thoſe little Difficulties incident to human Nature, without the leaſt Anxiety.

Such a Protheſis can ſcarce be called leſs than an ocular Revelation, not only ſhewing us how reaſonable it is to expect a future Life, but as it were, pointing out to us the Buſineſs of an Eternity, and what we may with the greateſt Confidence expect from the eternal Providence, dignifying our Natures with ſomething analogous to the Knowledge we attribute to Angels; from whence we ought to deſpiſe all the Viciſſitudes of adverſe Fortune, which make ſo many narrow-minded Mortals miſerable.

I am now, &c.

LETTER the NINTH.

Reflections, by Way of General Scolia, *of Consequences relating to the Immortality of the Soul, and concerning Infinity and Eternity.*

SIR,

THIS my laſt Letter to you, I mean my final aſtronomical one, I propoſe as a *General Scolia* to the reſt, the principle Matter being Reflections upon what is gone before, with ſome Concluſion naturally following or appendant to what has been already ſaid; but which, I could not in any other Place, ſo properly remark to you.

The Probability of the foregoing Conjectures, chiefly built upon very diſtant Obſervations, ſhew an apparent Neceſſity for ſome other kind of Doctrine permitted by Providence, to give Mankind a Knowledge of their Immortality and Dependance upon it, in the firſt Ages of the World.

And for the ſame Reaſon it evidently appears, that the ancient Philoſophers had it not in their Power to prove a ſupream *Being* and Director of all Things this Way.

And yet, as by a Sort of Inſtinct, or natural Reaſon, and Conſciouſneſs of a *good Principle*, we ſee how many noble Steps they made towards it, and was convinc'd at laſt of this *great Truth*, that ſince there was a *Mind* in ſo imperfect a Creature as Man, the *perfect Univerſe*, which comprehended all Things, could not poſſibly be without one ; and as Sir *Iſaac Newton* has juſtly obſerved in his *Principia*, " If every Par-" ticle of Space be *always*, and every individual Moment of Duration " *every where* ; ſurely the Maker and Lord of all Things, cannot be *never* " and *no where*."

To make manifeſt the infinite Empire and Agency of God, from celeſtial Motion, became the Taſk, but of very late Years ; and I can't help being of Opinion, that by means of theſe primary Bodies, only, we ſhall at length be able to trace the greater Circulations, and Laws of Nature, to their real original and fountain Head.

Theſe

These, were any thing wanting, befides the *Miracle ourfelves*, to convince us of a divine Origination, are all infallible Proofs, that the Univerfe is governed by an intelligent and all-powerful Being, whofe Exiftence is too nearly related to a felf-evident Truth to be more clearly demonftated, than it is manifeft of itfelf, both from the particular Laws of Nature, and the general Order of Things. An Argument which has been thought of no fmall Force, and well worth obferving in the Infancy of *Chriftianity*. *The invifible Things of God are clearly feen, being underflood by the Things that are made, even his eternal Power and Godhead.* Rom. i. 20.

But 'tis now high time to look back upon my Theory, and tell you it is a vain Suppofition, to imagine I fhall ever be able to convince every Reader, either of the Truth or Probability of what I have advanced to you : Mathematical Affiftance not being to be expected, where perhaps it has never been thought of ; and I allow you, it is much more reafonable to expect, that fifty Perfons will read thefe Letters without perceiving the Reafonablenefs of them, than that five fhould confider them with proper Judgment.

I muft ingenuoufly confefs to you, that nothing is wanting to convince me intirely of the Certainty of what I here advance by way of Conjecture to you. But this you muft only look upon as an happy Partiality, which generally attends all Authors, and always will be the chief Support of their tedious Labours. I affure you, I have neither Hopes nor Expectation, no, not the weak Breath of a Wifh, to be admitted a proper Judge of my own Works. But I fhall always take their Imperfection to be rather, (like my own Faults) to be too near me to be feen ; I therefore truft all to my Friend, and if I am fo fortunate as to excite his Approbation, I fhall think myfelf very happy in a very favourite Point ; which is, The advancing nothing which a rational Reader would willingly overlook, or be ignorant of.

But if I have been fo happy as to come fo near the Mark, as to border upon Truth, I believe you will allow me to carry my Conjectures a little further, and point out fome farther pleafing Confequences, which I begin to perceive may naturally follow.

Should it be granted, that the Creation may be circular or orbicular, I would next fuppofe, in the general Center of the whole an intelligent Principle, from whence proceeds that myftick and paternal Power, productive of all Life, Light, and the Infinity of Things.

Here the to-all extending Eye of Providence, within the Sphere of its Activity, and as omniprefently prefiding, feated in the Center of Infinity, I would imagine views all the Objects of his Power at once, and every Thing immediately direct, difpenfing inftantaneoufly its enlivening Influence,

to

to the remoteſt Regions every where all round. I know you'll ſay Aſtronomers are never to be ſatisfied, and I muſt own where there is ſo much
rational Entertainment for the human Mind, and ſo ſuitable to the
true Dignity of God, and moſt worthy of Man, it is not eaſy to
know where to ſtop in ſuch a Scene of Wonders.

Having, I ſay, once granted that all the Stars may move round one
common Center, I think it is very natural to one, who loves to purſue
Nature as far as we may, to enquire what moſt likely may be in that
Center; for ſince we muſt allow it to be far ſuperior to any other Point
of Situation in the known Univerſe, it is highly probable, there may be
ſome one Body of ſiderial or earthy Subſtance ſeated there, where the divine
Preſence, or ſome corporeal Agent, full of all Virtues and Perfections,
more immediately preſides over his own Creation. And here this primary Agent of the omnipotent and eternal Being, may ſit enthroned, as
in the *Primum Mobile* of Nature, acting in Concert with the eternal
Will. To this common Center of Gravitation, which may be ſuppoſed
to attract all Vertues, and repel all Vice, all Beings as to Perfection
may tend; and from hence all Bodies firſt derive their Spring of Action,
and are directed in their various Motions.

Thus in the *Focus*, or Center of Creation, I would willingly introduce
a primitive Fountain, perpetually overflowing with divine Grace, from
whence all the Laws of Nature have their Origin, and this I think would
reduce the whole Univerſe into regular Order and juſt Harmony, and
at the ſame time, inlarge our Ideas of the divine Indulgence, open our
Proſpect into Nature's fair Vineyard, the vaſt Field of all our future
Inheritance.

But what this central Body really is, I ſhall not here preſume to ſay,
yet I can't help obſerving it muſt of Neceſſity, if the Creation is real and
not merely Ideal, be either a Globe of Fire ſuperior to the Sun, or
otherwiſe a vaſt terraqueous or terreſtial Sphere, ſurrounded with an
Æther like our Earth, but more refined, tranſparent and ſerene. Which
of theſe is moſt probable, I ſhall leave undetermined, and muſt acknowledge at the ſame time, my Notions here are ſo imperfect, I hardly dare
conjecture. 'Tis true, I have ventur'd to think it may be one of theſe,
and ſince ſo glorious a Situation can hardly be ſuppoſed without its proper Inhabitants, 'tis moſt natural to conclude it may be the latter. In
the firſt Caſe, beſides our having no Idea of Beings exiſting in Fire, it
would not, notwithſtanding its Diſtance, be ſo eaſy to account for its being
inviſible; and ſince the Luſtre of the Stars are all innate, they could
receive no Benefit from it, and conſequently ſuch a Nature as a ſolar Com-
poſition, muſt in this Place be render'd uſeleſs; but in the latter Sup-

poſition.

pofition of its being a dark Body, we have no Difficulty attending us, having feveral Inftances of like Bodies, moving round an opaque one. Now when we confider, that all thofe radient Globes, which adorn the Skies, thofe bright ætherial Sparks of elemental Fire, thick ftrewed like Seeds of Light all round our Hemifphere, are each to us the Embrio of a glorious Sun ; how awful and ftupendious muft that Region be, where all their Beams unite and make one inconceivable eternal Day ?

Though the Deity, fays a learned Writer " be effentially prefent thro' " all the Immenfity of Space, there is one Part of it in which he difco- " vers himfelf in a moft tranfcendent and vifible Glory. This is that Place " which is mark'd out in Scripture, under the different Appellations " of PARADICE; *the third Heaven*; *the Throne of* GOD, *and the Habi-* " *tation of his Glory.*"

THIS continues the fame Author, is " that Prefence of God, which " fome of the Divines call his glorious, and others his majeftick Pre- " fence."

It is here, and here only, as in the Center of his infinite Creations, where he refides in a fenfible Magnificence, and in the midft of thofe Splendors, which can Effect the Imagination of his Creatures ; and though the moft facred and fupreme Divinity be allowed as effentially prefent in all other Places as well as in this, as being a BEING whofe Center is every where, and Circumference no where; yet it is here only, or in fuch Senforium of his Unity, where he manifefts his corporeal Agency, as in the Foci of his infinite Empire over all created Beings. It is to this majeftick Prefence of GOD, we may apply thofe beautiful Expreffions of Scripture, " *Behold even to the Moon and it fhineth not* ; *yea the Stars* " *are not pure in his Sight.*"

" The Light of the Sun, and all the Glories of the World, on which " we live, are but as weak and fickly Glimmerings, or rather Dark- " nefs it felf, in Comparifon of thofe Splendors, which encompafs this " Throne of GOD."

Here Heav'ns wide Realms an endlefs Scene difplays,
And Floods of Glory thro' its Portals blaze ;
The Sun himfelf loft in fuperior Light,
No more renews the Day, or drives away the Night :
The Moon, the Stars, and Planets difappear,
And Nature fix't makes one eternal Year.

Here and here alone center'd in the Realms of inexpreffible Glory, we juftly may imagine that primogenial Globe or Sphere of all Perfections,

fubject

subject to the Extreams of neither Cold nor Heat, of eternal Temperance and Duration. Here we may not irrationally suppose the Vertues of the meritorious are at last rewarded and received into the full Possession of every Happiness, and to perfect Joy. The final and immortal State ordain'd for such human Beings, as have passed this Vortex of Probation thro' all the Degrees of human Nature with the supream Applause.

What vast room is here, for infinite Power and Wisdom to act in, and that so visibly takes Delight to bless all his Beings with his Bounty. And endless as his Prescience, Attributes, and Goodness, are undoubtedly all those natural and apparent Joys with which he manifests his Love to all his Creatures, a Multiplicity of Objects not to be enumerated. For wheresoever we turn our Eyes, and follow with our Reason, we may meet with Worlds of all Formations, suited no doubt to all Natures, Tastes, and Tempers, and every Class of Beings.

Here a Groupe of Worlds, all Vallies, Lakes, and Rivers, adorn'd with Mountains, Woods, and Lawns, Cascades and natural Fountains; there Worlds all fertile Islands, cover'd with Woods, perhaps upon a common Sea, and fill'd with Grottoes and romantick Caves. This Way, Worlds all Earth, with vast extensive Lawns and Vistoes, bounded with perpetual Greens, and intersperfed with Groves and Wildernesses, full of all Varieties of Fruits and Flowers. That World subsisting perhaps by soft Rains, this by daily Dews, and Vapours; and a third by a central, subtle Moisture, arising like an Effluvia, through the Pores and Veins of the Earth, and exhaling or absorbing as the Season varies to answer Nature's Calls. Round some perhaps, so dense an Atmosphere, that the Inhabitants may fly from Place to Place, or be drawn through the Air in winged Chariots, and even sleep upon the Waves with Safety; round others possibly, so thin a fluid, that the Arts of Navigation may be totally unknown to it, and look'd upon as impracticable and absurd, as Chariot flying may be here with us; and some where not improbably, superior Beings to the human, may reside, and Man may be of a very inferior Class; the second, third, or fourth perhaps, and scarce allow'd to be a rational Creature. Worlds, with various Moons we know of already; Worlds, with Stars and Comets only, we equally can prove is very probable; and that there may be Worlds with various Suns, is not impossible. And hence it is obvious, that there may not be a Scene of Joy, which Poetry can paint, or Religion promise; but somewhere in the Universe it is prepared for the meritorious Part of Mankind. Thus all Infinity is full of States of Bliss; Angelic Choirs, Regions of Heroes, and Realms of Demi-Gods; Elysian Fields, Pindaric Shades, and Myriads of inchanting Mansions,

M.   not

not to be conceived either by Philofophy or Fancy, affifted by the ftrongeft Genius and warmeft Imagination.

All harmonioufly crowded and provided with every Object of Beatitude, that Friendfhip, Love, or Society can infpire, the Mufes or the Graces Frame; and all as permanent and perfect, that is deftin'd to a Duration, fuited to the Nature of their Exiftence and Degree of Cognifance; for as a very learned Writer obferves upon this fame Subject;

" How can we tell, but that there may be above us Beings of greater
" Powers, and more perfect Intellects, and capable of mighty Things,
" which yet may have corporeal Vehicles as we have, but *finer* and
" *invifible?* Nay, who knows, but that there may be even of thefe
" many *Orders*, rifing in Dignity of Nature, and Amplitude of Power,
" one above another? It is no Way below the Philofophy of thefe Times,
" which feems to delight in inlarging the Capacities of Matter, to affert
" the Poffibility of this."

From thefe amazing Ideas of Space in general, and from the particular Diftance of the Stars, which feparates as it were, one Syftem of Bodies from another, and by fo prodigious an extent, as fcarce to be fuppos'd a temporal Tafk. I think it naturally follows, had we no other Way to prove it, or any other Reafon to believe it, that the Soul muft of Neceffity be immaterial; for as this Space feems fo impaffible to Matter, as not to be undertaken and performed without the Lofs of Ages, in a State only of Tranfmigration, we may well imagine, that Change of Place is not effected this Way, but by fome other Vertue or Property, more immediate, if not inftantaneous.

I own next to *Annihilation* is the State of Oblivion, and this Way we may folve all Difficulties with regard to our being fenfible of fuch a Lofs of Exiftence; but if we allow the Soul to be immaterial, it no longer has any thing to do with Space, but as operating by Reflection only, or the Faculty of Thinking; it may be like the Imagination where it pleafes in a Moment.

Objects of the Mind abftracted from the Senfes of the Body, has no real or comparative Magnitude; that is, I would fay, an Inch, a Foot, a Yard, a Mile, or a Million of Miles are all equally indefinite, and is thus prov'd; every finite Line is formed of an infinite Number of Points, and no finite Line can be folv'd into more. Thus if you will allow me the Expreffion, the Mind being magnified as all Objects are diminifhed, what feems impracticable in the natural State of Things, in an Ideal one, becomes very poffible; that is, to make myfelf more intelligible, though we can hardly conceive, how any Being can pafs from *Syrius* to the Sun, by natural Laws in their proper State, yet if proportionally reduced by a

new

PLATE. XXXI.

new Modification of Ideas, to the Bignefs of a Ball 6 Feet Diameter, and to be only 680 Miles afunder; the Thing is very comprehenfive and eafy.

Hence Vifion, Light, and Electrical Virtue, feem to be propagated with fuch Velocity, that nothing but God can poffible be the Vehicle; and hence we may juftly fay with St. *Paul, Acts* xvii. 28. *In him we live, in him we move, in him we have our Being.*

It will further appear, from the foregoing Letters, that all the Stars and planetary Bodies within the finite View, are altogether but a very minute Part of the whole rational Creation; I mean that vaft collective Body of habitable Beings, which I have endeavoured to demonftrate, are all govern'd by the fame Laws, though varioufly revolving round one common Center, in which Center we may not impertinently venture to fuppofe the prime Agent of our Natures; or otherwife, the moft perfect of all created Beings, illimitable in his Ideas and Faculties of Senfation particularly prefide.

> But tho' paft all diffus'd, without a Shore
> His Effence; *local* is his Throne, (as meet)
> To gather the difperft, (as Standards call
> The lifted from afar) to fix a Point;
> A central Point, collective of his Suns,
> Since finite ev'ry Nature, but his own.   Dr. *Young*.

And farther fince without any Impiety; fince as the Creation is, fo is the Creator alfo magnified, we may conclude in Confequence of an Infinity, and an infinite all-active Power; that as the vifible Creation is fuppofed to be full of fiderial Syftems and planetary Worlds, fo on, in like fimilar Manner, the endlefs Immenfity is an unlimited Plenum of Creations not unlike the known Univerfe. See *Plate* XXXI. which you may if you pleafe, call a partial View of Immenfity, or without much Impropriety perhaps, a finite View of Infinity, and all thefe together, probably diverfified; as at A, B and C. in *Plate* XXXII. which reprefents their Sections, if all may be a proper Term for an infinite or indefinite Number, we may juftly imagine to be the Object of that incomprehenfible Being, which alone and in himfelf comprehends and conftitutes fupreme Perfection.

That this in all Probability may be the real Cafe, is in fome Degree made evident by the many cloudy Spots, juft perceivable by us, as far without our ftarry Regions, in which tho' vifibly luminous Spaces, no one Star or particular conftituent Body can poffibly be diftinguifhed; thofe in

all

177

all likelyhood may be external Creation, bordering upon the known one, too remote for even our Telefcopes to reach.

With the raptur'd Poet may we not juftly fay

> O, what a Root! O what a Branch is here!
> O what a Father! what a Family!
> Worlds! Syftems! and Creations!

And in Confequence of this

> In an Eternity, what Scenes fhall ftrike?
> Adventures thicken? Novelties furprize?
> What Webs of Wonder fhall unravel there?
>
> *Night Thoughts.*

So many varied Seats where every Element may have its proper Beings and all adapted to partake of every thing fuited to their Natures, argue fuch Maturity of Wifdom, and the vaft Production fuch myfterious Power; 'tis hardly poffible for Mortals not to fee divine Intelligence prefide, and that every Being fomewhere muft be happy.

A Univerfe fo well defigned, and fill'd with fuch an endlefs Structure of material Beings, and all the Refult of Prefcience and infinite reflected Reafon, flowing from a Mind all perfect, full of all Ideas, could never be defigned in vain; and tho' our narrow Bounds of Reafon limited, by finite Senfes, cannot directly fee the Confequence dependant on a Sequel, yet from what we do fee, great Room we have to hope the next Stage of Exiftence will be more lafting and more perfect; and it is highly probable, the nobleft Suggeftion of the moft luxuriant Fancy may fall infinitely fhort of what we are defigned for.

But here, even in this World, are Joys which our Ideas of Heaven can fcarce exceed, and if Imperfection appear thus lovely, what muft Perfection be, and what may we not expect and hope for, by a meritorious Acquiefcence in Providence, under the Direction, Indulgence, and Protection of infinite Wifdom and Goodnefs, who manifeftly defigns perfect Felicity, as the Reward of Virtue in all his Creatures, and will at proper Periods anfwer all our Wifhes in fome predeftined World.

All this the vaft apparent Provifion in the ftarry Manfions, feem to promife: What ought we then not to do, to preferve our natural Birthright to it and to merit fuch Inheritance, which alas we think created all to gratify alone, a Race of vain-glorious gigantick Beings, while they are confined to this World, chained like fo many Atoms to a Grain of Sand.

I am, &c.